THE QUATERNARY PERIOD

Years B P	EPOCH	GLACIALS & North America	Interglacials Europe
	Holocene	(The present interglacial)	
10,000			
	Pleistocene	WISCONSIN	WÜRM
		Sangamon	Riss–Würm
		ILLINOIAN	RISS
		Yarmouth	Mindel–Riss
		KANSAN	MINDEL
		Aftonian	Gunz–Mindel
		NEBRASKAN	GUNZ
			Donau–Gunz
		?	DONAU
		(Pre – glacial)	
? 2,000,000			

Biogeography

BIOGEOGRAPHY

E. C. PIELOU

Dalhousie University
Halifax, Nova Scotia

A Wiley-Interscience Publication

JOHN WILEY & SONS

New York • Chichester • Brisbane • Toronto

Library of Congress Cataloging in Publication Data:

Pielou, E C 1924–
    Biogeography.

    "A Wiley-Interscience publication."
    Bibliography: p.
    Includes index.
    1.   Geographical distribution of animals and plants.
2.   Evolution. I. Title.

QH84.P53   574.9   79-13306
ISBN 0-471-05845-9

Printed in the United States of America

10 9 8 7 6 5 4

PREFACE

The word "biogeography" denotes a subject of enormous breadth, so enormous that attempts are rarely made to deal with more than a few facets of it in one book. It can be subdivided on the basis of regions, organisms, or concepts. For example, marine biogeography has been treated by Ekman (1953) and Briggs (1974); vertebrate zoogeography by Darlington (1957); phytogeography from an ecological point of view by Dansereau (1957); phytogeography from a taxonomic point of view by Good (1974); ecological biogeography by Watts (1971); paleobiogeography by Simpson (1953), Kurtén (1971), and Valentine (1973); island biogeography by MacArthur and Wilson (1967) and Carlquist (1974). In addition there are symposium volumes devoted to the biogeography of various regions such as the Pacific Basin (Gressit, 1963), the Bering Strait region (Hopkins, 1967), the North Atlantic (Löve and Löve, 1963) and the southern continents (Keast, Erk, and Glass, 1972). And these are only a few representative examples.

This book attempts (probably rashly) to cover the whole field, at a more advanced level than that of a survey for beginners. It is intended for senior undergraduates, and for interested biologists and earth scientists of all kinds. My purpose in writing it is to make students aware of the tremendous range and diversity of biogeography. The subject is so large that a research worker cannot hope to be expert in more than one or two branches of it, but at the same time cannot afford to be ignorant of the multitude of other branches. In biogeography perhaps more than in any other subject, a holistic approach is necessary. The solution of any biogeographic problem, no matter how small, requires the following of a great many varied lines of inquiry.

Although no one person can weigh all the diverse bits of evidence bearing on a particular problem, he should at least be conscious of their existence and relevance. It therefore seems worth while to bring together in one book a set of "samples" of biogeography's many branches. The chapters can be treated as a series of essays but there is much cross-referencing. The need for cross-referencing brought home to me, as I wrote, the degree to which the threads of the subject are interwoven. The desirability of ample cross-referencing is the justification I offer for presuming to write on so broad a range of topics.

I am very grateful to Peter Raven, Director of the Missouri Botanical Garden, for wise and useful advice (I took most of it). I am also indebted to Marjorie Willison of Halifax for considerable help in gathering and collating material, and to Terry Collins of Halifax who did the drawings and most of the diagrams.

E. C. PIELOU

Halifax, Nova Scotia
May 1979

CONTENTS

Biogeography

INTRODUCTION

A book, or a series of lectures, is a one-dimensional stream of words. This fact constrains all scientific writers, whatever their topic. Any branch of science has, in reality, a multidimensional framework of concepts, and the task of projecting this framework onto a single line in such a way that the inevitable distortions do not obliterate the intricacies of the structure is the hardest part of scientific writing. In the case of so-called interdisciplinary subjects (and biogeography is preeminently one of them), the difficulties are compounded. Where best to begin the subject, and how best to divide it into compartments, are debatable problems with no unique answers.

The work of a biogeographer consists in observing, recording, and explaining the geographic ranges of all living things. These ranges are not static. Their salient feature is that they are continually expanding, contracting, fragmenting, and coalescing. Moreover, these changes are occurring over a whole hierarchy of different time scales. Obviously the causes and effects of short-term changes, those taking a few decades, differ fundamentally from those taking tens or hundreds of millions of years.

Changes in the geographic range of a biological species-population, just as much as changes in the morphology and physiology of its member individuals, are all part of the evolution of that species. Range-changes influence, and are influenced by, changes in a species' biology. Thus biogeography should not be thought of as a fringe subject, an optional supplement to a biologist's training for students whose timetables have room for it. It is, in truth, right at the heart of biology, an integral part of the study of evolution.

The current upsurge of interest in biogeography follows, and is part of, the scientific revolution undergone by the earth sciences in the late 1960's. This revolution consists in the reversal of opinion of the great majority of geophysicists on the reality of continental drift. Before it, most geophysicists regarded drift as physically impossible (Wilson, 1963a). The switch in consensus coincided with the development and widespread acceptance of the theory of plate tectonics, according to which the earth's crust is formed of a number of thin, rigid plates that are constantly in motion relative to each other. They are forced apart, along various lines of separation, by the process of sea-floor spreading. With the demonstration by geophysicists that this process can and does occur, a belief in continental drift suddenly gained respectability. Before the revolution, only a few biologists were tough-minded enough to maintain, in the teeth of opposition, that drift *must* occur since it is the only reasonable explanation of a great many well-known biogeographic patterns. Since the revolution, they have been joined by most of their more timid colleagues, who had previously allowed their opinions to be formed for them by geophysicists. Perhaps the moral has now been perceived: biological evidence for continental drift is as weighty as geophysical evidence. Biologists who denied this, and who deferred to geophysicists on the subject, were undervaluing their science.

The changes in biogeographic range-patterns caused by plate movements are, of course, long-term changes, and are only one of the scales of change discussed in this book. A synopsis of its contents will show how I have dismantled the multidimensional structure of the science of biogeography so as to present it in a sequence of chapters.

To begin (Chapter 1) we consider modes of biogeographic classification: the subdivision of the present-day earth into faunal and floral realms and provinces.

The biogeographic consequences of plate tectonics and modern interactions between biogeography and geophysics are the subjects considered in Chapter 2.

Chapter 3 stresses the evolutionary consequences of biogeographic change, and the conclusions about evolutionary processes that may be derived from biogeographic evidence. These are topics of heated controversy.

In Chapter 4 we concentrate on a much shorter time interval in earth history than that covered in Chapter 2 and discuss the biogeographic effects of the last ice age.

Chapter 5 describes aspects of biogeography that are peculiar to marine organisms. The continuity of the world ocean, and the effects of ocean currents, obviously lead to some fundamental differences between terrestrial and marine biogeography.

Chapter 6 discusses island biogeography, the subject founded by MacArthur and Wilson (1963). It forms a distinct field of study. Large-scale events in earth history (such as the opening of the Atlantic Ocean) and large-scale units of the earth's surface (the continents and oceans) are unique and have no replicas. Small islands, in contrast, are so numerous that biological events occurring in them can be thought of, in the statistical sense, as samples from larger, parent populations. This contrast sets "island" theory in a class by itself.

Chapter 7 deals with geoecology. This includes (among other things) a study of the recurrence, in geographically similar areas, of ecologically similar communities composed of taxonomically different members.

Chapter 8, on biological dispersal and diffusion, brings together much that has already appeared, in scattered form, in earlier chapters.

Chapter 9, likewise, is a synthesizing chapter. Biogeographic range-disjunctions are always thought-provoking and have many different causes. Examples are described, in a wide array of different contexts, throughout the earlier chapters of the book. Chapter 9 brings them together in brief, summary form.

Chapter 10 deals with geographic patterns in polymorphic species, and geographic variations in polyploidy.

Deciding which chapter was the appropriate home for a topic was often difficult. Many topics have an equal claim to inclusion in two, three, or more chapters and the choice was arbitrary. This merely illustrates the multidimensional linkages that characterize the framework of the body of knowledge known as "biogeography."

Statistical and mathematical reasoning and methods are gradually seeping into biogeography. The few sections of the book that presuppose some knowledge of mathematics and statistics are marked with asterisks. The reader who wishes to omit these sections can be assured that without them the book still forms a connected whole. But to skip them is to miss a taste of the direction in which biogeography seems most likely to advance.

In Chapters 2, 3, and 4 are repeated references to geological time intervals of different ranks (eras, periods, epochs, and smaller units), and to the times, in millions of years (m.y.) before the present (BP) of various events in earth history. As an aid to those whose background is chiefly in neobiology, two tables are printed on the endpapers. One shows the major divisions of geological time since the start of the Mesozoic era about 230 m.y. ago. The other, on a larger scale, shows the estimated dates and durations of the various glacial episodes of the Quaternary ice age.

Chapter One

THE BIOGEOGRAPHIC SUBDIVISIONS OF THE EARTH

Biogeography is concerned with plant and animal species and also, from the very nature of the science, with entities such as biogeographic realms, regions, and provinces. The biogeographic classification of the modern earth is therefore the first topic to consider in a study of biogeography, and is the subject of this chapter.

There are three sections. The first discusses the units into which the whole earth has been divided, on the basis of accumulated knowledge of floras and faunas, carefully, but subjectively, weighed and considered. Section 2 describes one of the many possible ways of performing an objective classification, using multivariate statistical methods, of a comparatively small area; that is, an area small relative to the whole earth. In the example given, Australia is the area classified. Section 3 deals with methods of delimiting biogeographic provinces in a rather special case, when the region to be subdivided can be treated for all practical purposes as one-dimensional. This is true of a shoreline, inhabited along a very narrow strip by a biota of strictly littoral organisms.

1. BIOGEOGRAPHIC REALMS, REGIONS, AND PROVINCES

A "biological" subdivision of the earth's surface can take account either of the terrestrial or the marine biosphere. Obviously, the results are very different. In a terrestrial subdivision, boundary lines are drawn through the oceans which are treated as "lifeless" for the purpose at hand; and *mutatis mutandis* for a marine subdivision. In what follows, we consider the terrestrial biosphere first.

Further, a "biological" subdivision of the earth's surface can be based on taxonomic or ecological criteria. A biogeographic subdivision, as the term is ordinarily used, is one using taxonomic criteria. Of course no biologist can afford to ignore the ecological differences between different areas, and a list of the world's ecological units of highest rank, the biomes, is given in a later subsection.

A Taxonomic Classification of the Terrestrial Biosphere

A "perfect" biogeographic classification of the terrestrial world is, of course, an unattainable ideal. Disagreements over the ranks to be assigned to the recognized units, and over the exact locations of their boundaries, are inevitable. Somewhat different subdivisions are obtained depending on whether a classification is based on the floras or the faunas of the different areas.

The following classification (Table 1.1) and map (Figure 1.1) will serve as bases for the following discussion.

Table 1.1 is the classification of Schmidt (1954; see Sylvester-Bradley, 1971). As may be seen, units of four different ranks are recognized: the realm, the region, the subregion, and the province. This is a zoogeographic classification. In order to arrive at a map of the world's biogeographic divisions which, as far as possible, gives equal emphasis to animals and plants, I have combined Schmidt's provinces, as shown by the boxes in Table 1.1, into eight "biogeographic regions" which are mapped in Figure 1.1.

Examination of the table and the map naturally leads one to ask how, and why, the subdivisions were arrived at. To say that the flora and fauna of any one region exhibit uniformities that set it off from all other regions is to beg the question. It would, in any case, be truer to say that biogeographic units are marked more by the differences between them than by the resemblances within them. It is by their boundaries that they are best defined. These boundaries could be more precisely described as transition zones. Some are narrow, others wide; some are clear and definite, others ill-defined. They merit careful study in their own right and we return to the topic below.

First it is worth considering how we might characterize the different regions shown in Figure 1.1. One way is to list the families of mammals that are endemic to each region. (Mammals are useful for illustrative

Table 1.1 A zoogeographical classification of the world (after Schmidt, 1954). The "biogeographic regions" mapped in Figure 1.1 are not Schmidt's "zoogeographic regions" tabulated below. The regions in the map have been constructed by combining Schmidt's provinces as shown by the boxes in the rightmost column

Realm	Region	Subregion	Province
Arctogaean	Paleotropical	Oriental	Indian. Ceylonese. Indo-Chinese. Malayan. Celebesian[a]
		Ethiopian	West African. South African
		Malagasy	Seychellian, Madagascan, Mascarene
	Holarctic	Palearctic	European. Siberian. Manchurian. Tibetan. Eremian.[b] Mediterranean
		Arctic	Old World Arctic
			New World Arctic
		Nearctic	Canadian. Appalachian. Western American. Sonoran
Neogaean	Neotropical	Caribbean[c]	Central American[a] West Indian
		Neotropical	Amazonian. East Brazilian. Chilean

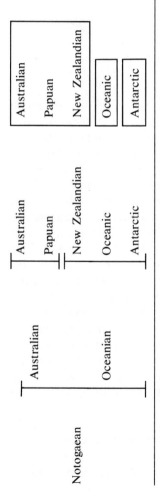

Notogaean — Australian, Oceanian

Australian, Papuan, New Zealandian, Oceanic, Antarctic

Australian, Papuan, New Zealandian, Oceanic, Antarctic

[a] More often treated as a between-realm transition zone than as a province.

[b] The Central Asian desert region.

[c] More often treated as part of the Neotropical than of the Holarctic region.

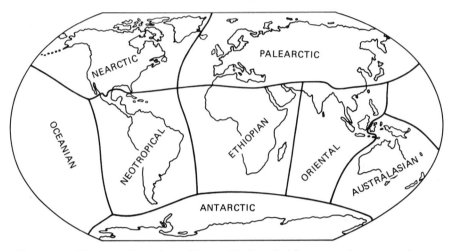

Figure 1.1 Biogeographic regions of the world. Compiled from several sources and combining zoogeographical and phytogeographical systems.

purposes since all biologists, whatever their specialty, are mammals themselves and presumably familiar with many, at least, of their own kind from visits to zoological parks.) Excluding marine mammals and bats, there are 89 mammal families in the world fauna (George, 1964). Of these, only three are cosmopolitan, that is, present in every region except the Antarctic. These are the Old World rats and mice (family Muridae; order Rodentia), the rabbits and hares (Leporidae; Lagomorpha) and the dogs and wolves (Canidae; Carnivora). These animals owe their present cosmopolitan distribution to human agency and if their present large ranges are regarded as "unnatural," then the only truly cosmopolitan terrestrial mammal families are the bats (order Chiroptera) which, because they can fly, are a special case excluded from consideration here. (Also excluded, for a similar reason, are the Hominidae.)

Of the remaining 86 families, 50 are endemic to a single region and the remaining 36 occur in two or more regions. Thus endemic mammal families serve to typify different regions and a list of a region's endemic mammals helps one to visualize the region. Table 1.2 gives these lists for the six regions that have endemic families. Two regions (Oceania and Antarctica) lack them, and those that have them differ greatly in the number they contain.

As remarked above, a region is often better defined in terms of the boundaries, or transition zones, that separate it from other regions than in terms of its own biota. These transition zones are of different kinds. For example, as a glance at Figure 1.1 shows, the Australasian region is separated from all others by salt water whereas the Ethiopian and Palearctic regions are separated by desert.

Different kinds of barriers differ from each other in their imperviousness, that is, in the extent to which they prevent any interchange of biotas across them; also, they affect different kinds of organisms differently. Thus even a narrow saltwater strait is an impassable barrier to freshwater fishes, but plants with wind-dispersed seeds can cross easily. Another example: a stretch of frozen sea is a barrier to small mammals but not to large ones. This explains the presence in Newfoundland of such animals as red fox (Canidae), beaver (Castoridae), caribou (Cervidae), and the snowshoe and arctic hares (Leporidae), and the absence of the following small mammals which are abundant on the adjacent Canadian mainland: moles (Talpidae), shrews (Soridae), squirrels, marmots and chipmunks (Sciuridae), and deer mice, lemmings, and voles (Cricetidae) (Burt and Grossenheider, 1961). Only the larger animals were able to cross the sea ice and occupy the island when the Pleistocene ice sheets melted.

The degree of contrast between the regions on either side of a boundary depends, also, on the age and history of the boundary. Boundaries form and vanish as land masses drift apart and come together. Similarly, they form and vanish with changing climates; as a drifting continent moves through a series of latitude zones, its climate becomes cooler or warmer, moister or more arid. Thus a given tract of land may be a desert in one geological epoch and forest covered in another. These matters are the theme of Chapter 2, but it is worth emphasizing here that the boundaries between biogeographic units, and hence the units themselves, are continuously changing. The longer a barrier has existed, the older will be the regions on either side of it, and the higher the taxonomic rank of their endemic organisms.

Barriers change in kind and in imperviousness. For example, until the middle Pliocene (about 6 m.y. BP) the Nearctic and Neotropical regions were separated by salt water; the barrier between them became much less difficult to cross with the formation of a land link between them, the Isthmus of Panama. Another example: until the middle Eocene, some 45 m.y. BP, India was separated from Asia by sea; the drifting of a crustal plate then brought India (previously an island) into contact with Asia, and a chain of mountains was forced up along the line of collision. As a result, the barrier between what are now the Palearctic region and a large segment of the Oriental region changed from salt water to mountains.

At the present day, the boundaries between geographic regions tend to be either saltwater barriers, or what Darlington (1957) has called zonal-climatic barriers. That is, several of them lie roughly along the parallel of latitude separating the tropics from the north temperate zone where,

Table 1.2 Families[a] of mammals endemic to the different biogeographic regions.[b] (In parentheses below the name of each region is the number of families endemic to it.)

Palearctic (2)	Nearctic (4)	Ethiopean[c] (15)	Neotropical (17)	Oriental (4)	Australasian (8)
RODENTIA Spalacidae (mole rat) Selevenidae	RODENTIA Geomyidae (pocket gopher) Heteromyidae (pocket mouse) Aplodontidae (sewellel) ARTIODACTYLA Antilocapridae (pronghorn)	RODENTIA 5 families ARTIODACTYLA Giraffidae (giraffe) Hippopotamidae (hippopotamus) TUBULIDENTATA Orycteropidae (aardvark) INSECTIVORA 3 families of shrews --- Tenrecidae (tenrec)	RODENTIA 11 families including guinea pig and capybara INSECTIVORA Solenodontidae (solenodon) PRIMATES Cebidae (New World monkeys) Callithricidae (marmoset)	RODENTIA Platacanthomyidae (spiny dormouse) PRIMATES Tupaiidae (tree shrew) Tarsiidae (tarsier) DERMOPTERA Cynocephalidae (colugo)	MARSUPIALIA Dasyuridae (Tasmanian wolf, now extinct) Peramelidae (bandicoot) Phascolomidae (wombat) Phalangeridae (phalanger) Notoryctidae (marsupial mole) Macropodidae (kangaroo. wallaby)

PRIMATES
Lemuridae
(lemur)
Indridae
(woolly lemur)
Daubentonidae
(aye aye)

MARSUPIALIA
Caenolestidae
(marsupial mouse)
EDENTATA
Myrmecophagidae
(South American
anteater)
Bradypodidae
(sloth)

MONOTREMATA
Tachyglossidae
(spiny anteater)
Ornithorhynchidae
(platypus)

[a]Families are grouped by orders, and the names of the orders are in capitals. Where possible, the popular name of a member of each family is given in parentheses below the family name.
[b]The regions are as shown in Figure 1.1.
[c]Families restricted to Madagascar are shown below the dashed line in this column.

Table 1.3 Transition zones between biogeographic regions (data from Darlington, 1957)

Transition zone	Location°	Nature of barrier
Nearctic-Palearctic ? ←——→	Bering Strait Norwegian Sea	Water gaps and cold climate
Nearctic-Neotropical birds ——→ reptiles and mammals ←—	Isthmus of Panama	Narrow land link replacing earlier water gap; arid areas in Mexico (i.e., zonal climatic barrier)
Ethiopian-Palearctic ——→	Sahara Desert	Arid desert (less arid in past) (i.e., zonal climatic barrier)
Ethiopian-Oriental ? ←——→	Southwest Asia and Arabian Peninsula	Arid lands (i.e., zonal climatic barrier)
Oriental-Palearctic ——→	Himalayas and eastward extensions	High mountains
Oriental-Australasian ——→	"Wallacea"	Water gaps

owing to the zonal nature of the world's climates, deserts tend to form. Table 1.3 summarizes the properties of those of the interregional boundaries that are not wide tracts of ocean. The arrows under the zones' names show the current direction in which such dispersal as there is most often takes place.

Another property of these boundaries, or rather of these transition zones, is that they tend to have depauperate faunas (Darlington, 1957). This is true of all but the zone separating the Oriental and Palearctic regions. Apart from this one exception, the transition zones are what Darlington calls "subtraction-transition areas" in which are found only a few of the faunal elements from either side.

It is interesting to consider a particular example in detail, namely the transition zone between the Australasian and Oriental regions, and how the mammal fauna of each region thins out as one crosses the zone away from it. The following discussion is based on that in Darlington (1957) and is illustrated by Figure 1.2

Figure 1.2*a* is a bathymetric map showing the 200 m depth contour. Shallower waters (shown unshaded) have probably been dry land from time to time as a consequence of changes in sea level both eustatic and

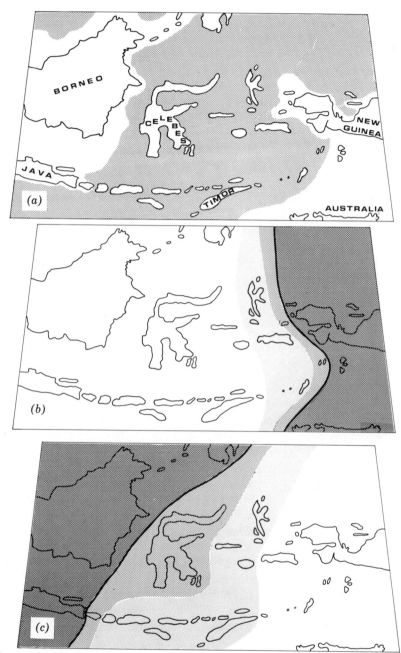

Figure 1.2 The densities of mammal species in Indonesia. (a) Bathymetric map. The contour is at 200 m. (b) Australasian mammals. The zones show areas whose fauna is, from right to left: wholly Australasian: with a few Australasian species; still fewer; none, except for two species of Phalanger. (c) Oriental mammals. The zones show areas whose fauna is, from left to right: wholly Oriental; a few Oriental species (shrews, civets, pigs, deer, monkeys); still fewer Oriental species, possibly all introduced; no Oriental species.

isostatic.* It is exceedingly unlikely, however, that a land bridge has ever crossed the whole area and linked the two regions.

Figure 1.2b shows the thinning out of the Australasian mammal fauna (details are given in the legend). The heavy line, to the east of which the fauna is wholly Australasian, is known as Weber's line. It is interesting to note parenthetically that Australasian mosquito species thin out westward in a pattern very like that of the mammals (Mattingly, 1962).

Figure 1.2c shows the thinning out of the Oriental mammal fauna. The heavy line, to the west of which the fauna is wholly Oriental, is Wallace's line.

The zone between the two lines is sometimes called Wallacea. The whole zone, rather than either of the lines, is the true boundary separating the Oriental and Australasian regions. The exact locations of the lines themselves is a matter of choice, since the true range limits of forest mammals of retiring habits are impossible to pin down precisely. In any case, Wallacea is a "subtraction" zone with few mammal species, and most of these are probably recent immigrants brought in by man, intentionally as pets or meat providers, or inadvertently as pests.

Classification of the Marine Biosphere

In subdividing the seas and oceans biogeographically, the same hierarchy of units used by terrestrial biologists is often used; that is, realms, regions, and provinces are recognized. But, as we shall see, while the two great marine realms differ fundamentally from each other, the lower-ranking units tend to be ill-defined.

The two realms are the *continental shelf* and the *deep ocean*. The difference between them is basic. The floor of the deep ocean has never been dry land. The floors of the so-called shelf seas are actually parts of the continental land masses that chance to be submerged, at present, under epicontinental seas. Changes in sea level, and crustal warping, cause the outlines of these shallow seas to vary. At some times (as we shall see in Chapter 2) they have been more extensive than they are now and shallow seas have covered parts of what are now continental interiors. At other times they have been less extensive and surfaces now submerged have been subaerial. This distinction is geophysical and a geophysicist usually places the boundary between the two realms at the

*These terms are defined on page 111, where sea level changes, their causes, and biogeographical consequences, are further discussed.

2000 m bathymetric contour, partway down the continental slope, which is, in fact, the zone separating the two realms.

The 2000 m contour has no special significance for living organisms, however, and for biological purposes it is more useful to take the 200 m contour which marks, approximately, the lower limit of useful light penetration, in which photosynthesis can proceed, as the boundary between the "shallow" and "deep" realms.

Subdivision of these two realms into regions and provinces can be done, but the boundaries are less easily definable than in the terrestrial case, and the contrasts between units are much slighter. Moreover, since the sea constitutes a three-dimensional living space, the subdivision into areas performed by biogeographers is at right angles to the subdivision into biomes (see below) performed by marine ecologists.

From the biogeographic point of view, as opposed to the ecological, the floor of the deep ocean is indivisible; it is the home of a single worldwide fauna (Hedgpeth, 1957). Disregarding the deep-sea benthos (which could more properly be regarded as a biome) the seas and oceans can be divided into regions and provinces as has been done by Briggs (1974), for example. But, as remarked above, the units are not clear-cut. This is because the world ocean, unlike the land, is continuous. With few exceptions, families of marine organisms are widespread or cosmopolitan, and one has to descend to genera or even species in the taxonomic hierarchy to find distinctions among marine faunas. Marine species and genera tend to have bigger ranges than their terrestrial counterparts. Also, latitudinal zonation is more frequently and more clearly exhibited by marine than by terrestrial organisms (Hewer, 1971). Marine regions, in fact, tend to coincide with ocean "water masses," bodies of water distinguished from each other by their physical characteristics, especially temperature (see page 140). No impenetrable barriers separate them, and their positions are not immovable. However, the closer one approaches to the land, the clearer the separation of the fauna into provinces becomes, since river mouths and sandy shoals are barriers to the spread of littoral species.

An Ecological Classification of the Earth

As has already been remarked, a classification of the earth's surface into ecologically defined units—the highest ranking are *biomes*—is not ordinarily regarded as a biogeographic classification. However, biogeographic studies can never ignore the existence, characteristics, and geographic pattern of the world's biomes. Grouping them into three sets, terrestrial, marine, and littoral, they are as follows:

Terrestrial biomes	Marine biomes	Littoral biomes
Tundra	Coral reef	Rocky shore
Coniferous forest	Deep sea benthos	Sandy shore
Deciduous forest	Pelagic plankton	Muddy shore and
Grassland	Pelagic nekton	saltmarsh
Savannah	Neuston	Shallow sea benthos
Tropical rain forest		
Chaparral		
Desert		
Freshwater		

This classification is not, of course, the only one possible. A world map showing ecological biomes needs to be printed on a large scale and in color if it is to be legible, since biomes are defined only by their ecological properties and may occupy a space made up of numerous small, scattered fragments. The units recognized in the vegetation-type maps found in most good geographical atlases are equivalent to biomes.

*2. BIOGEOGRAPHICAL CLASSIFICATION BY AN OBJECTIVE METHOD

The division of the world into biogeographic realms, regions, and provinces, as discussed in Section 1, is a subjective process. The result represents, approximately, the consensus of all biogeographers, whatever their specialty. Although many would disagree with the details of the map in Figure 1.1—they might rename some of the regions, combine or subdivide some of them, or adjust the boundaries slightly—all would probably find it an acceptable compromise.

It is often desired to subdivide a province even further, into lower-ranking units; often, too, one wishes to base the subdivision on a single taxon, for example, the birds or the mammals or the woody plants. To do this satisfactorily some objective method is required. The data on which a subdivision is to be based commonly consist of lists of species (of the taxon concerned) at each of a number of scattered sites. The aim is to partition the sites into a number of classes in such a way that sites within a class resemble each other, or sites in different classes differ from each other, where the words "resemble" and "differ" have been given precise, operational meanings. To complete a biogeographic subdivision, it is then necessary to draw boundary lines on a map of the site locations in such a way that the member sites of any one class are grouped in a single area, separated by boundaries from all other areas.

This last step can present difficulties. It may turn out that the boundaries form an unacceptably intricate pattern: that, for instance, one "area" appears as a number of islands within another area. If this happens, it suggests that the sites are too narrowly defined, and that ecological differences among them are masking their biogeographic affinities. To prevent this, one must ensure that each sample site be so defined as to include data from all ecological environments in its geographic neighborhood.

It could be argued, with some truth, that the process of defining "sample sites" by combining point locations so as to smooth out ecological differences is subjective.* This cannot be helped. No classification can be wholly objective. Another subjective choice is unavoidable: that of the method to be used in classifying the sites. There are an enormous number of possibilities. A study of methods of classifying multivariate data now constitutes a scientific discipline in its own right, one that finds plenty of application in ecology. Reviews of ecological classification procedures may be found in Cormack (1971), Pielou (1977a), and Williams (1971). Here it is possible to discuss only one of the methods, and to demonstrate its application.

The method is known as *divisive information analysis* and it has been used by Kikkawa and Pearse (1969) to divide Australia into biogeographic subareas on the basis of land bird distribution. Before considering their results, the method is described and applied to a deliberately simplified, artificial, illustrative example.

Suppose we have species lists from n sites and that a total of s species has been encountered. The species are named A, B, C, ... Let the number of sites containing the jth species ($j = 1, . . . , s$) be a_j; (thus $1 \leq a_j \leq n$, since the jth species must have been found in at least one and at most n sites; therefore $\Sigma a_j \geq n$).

Now define the *information content* or *site diversity* of the n sites as I, where

$$I = sn \log n - \sum_{j=1}^{s} [a_j \log a_j + (n - a_j) \log (n - a_j)].$$

The argument underlying the definition of I is given in Pielou (1977a). It suffices to say here that I is low if the n sites' species lists resemble each other, and high if they differ.

Now divide the sites on species A. That is, allot the sites to two classes: into class A+ put all the sites where species A occurs; and into class A− put the rest, that is, all the sites from which species A is absent. Now

*A more objective technique is discussed in another context in Chapter 7.

calculate the site diversities, say I_{A+} and I_{A-}, for these two classes treated separately.

Next put

$$\Delta I_A = I - (I_{A+} + I_{A-}).$$

Thus ΔI_A measures the loss of information content, or the *diversity drop*, that results from dividing the total set of sites into the classes A+ and A−. It is the difference between the site diversity of the total set and the sum of the site diversities of the two subsets.

Compute ΔI_B, ΔI_C, . . . , in the same way.

One can now see which dichotomy yields the largest diversity drop, and this fixes the first division to be made in classifying the data. The original data have now been dichotomously split in a way that ensures that the resemblance of sites within classes shall be large and between classes shall be small. Next, each class is itself divided by a repetition of the method; and so on. A subjective decision may be made on when to stop. Or an objective "stopping rule" can be chosen: for example, one might decide to carry the subdivision no further in any class whose site diversity fell below an arbitrarily chosen level; or, alternatively, to stop when the diversity drop resulting from a subdivision was less than a chosen amount.

A single step of the method is demonstrated in Figure 1.3 and Table 1.4. The imaginary island shown in the maps has been sampled at $n = 5$ sites (shown as squares) and the $s = 3$ species, A, B, and C, occur at these sites in the manner shown. Table 1.4 shows the site diversities of the total data, and of the three possible pairs of classes formed by subdivision on one of the species. It is seen from the final column that the largest diversity drop is obtained when the sites are divided on species A. Hence the best "geographic" subdivision of the island is that shown in Figure 1.3*b*.

A real example, from Kikkawa and Pearse (1969), is given in Figure 1.4. Shown is a subdivision of Australia based on the distribution of genera of land birds. The dendrogram resulting from the two-stage classification is shown below the map; notice that the style of line (solid or dashed) forming the verticals of the dendrogram matches the corresponding boundaries on the map. Thus the first division (solid line) is into areas $(F_1 + F_2)$ and $(F_3 + F_4)$, with 193 and 143 genera respectively. Of course some genera are common to both areas so the sum of the number of genera in each area exceeds the total, 199. Kikkawa and Pearse carried their classification to the stage where 10 areas (rather than the 4 shown here) were distinguished. They also did classifications based on species and on

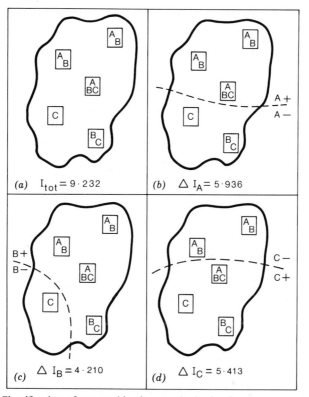

Figure 1.3 Classification of geographic sites on the basis of presences and absences of species. The "maps" show $n = 5$ sites containing $s = 3$ species, A, B, and C. (a) The undivided island. (b–d) Divisions of the island into two "faunal areas." The divisions are made on species A, B, and C respectively and below each is shown the resultant "diversity drops" ΔI_A, ΔI_B, and ΔI_C. Since the largest drop is ΔI_A, division on species A is best if the island is to be divided into two subareas (see Table 1.4 for computations).

superspecies of land birds, as well as on genera. Details are given in their paper.

*3. BIOGEOGRAPHICAL SUBDIVISION OF A LITTORAL BIOTA

Biogeographic Co-ranges

It is obviously easier to section a stretch of shore into biogeographic units than it is to subdivide a two-dimensional area. On a shore, the geographic

Table 1.4 The computations used for the areal classifications of an "island" shown in Figure 1.3

Class	Number of sites n	Number of species s	$\mathbf{a}_j{}^a$	Diversity[b] I	Diversity drop ΔI
Total	5	3	3, 4, 3	$I_{\text{tot}} = 9.232$	
A+	3	3	3, 3, 1	$I_{A+} = 1.910$	$\Delta I_A = 5.936$
A−	2	2	1, 2	$I_{A-} = 1.386$	
B+	4	3	3, 4, 2	$I_{B+} = 5.022$	$\Delta I_B = 4.210$
B−	1	1	1	$I_{B-} = 0$	
C+	3	3	1, 2, 3	$I_{C+} = 3.819$	$\Delta I_C = 5.413$
C−	2	2	2, 2	$I_{C-} = 0$	

$^a\mathbf{a}_j$ is the vector whose elements are the numbers of sites containing species A, B, and C respectively.
^{b}Natural logs were used in the calculations.

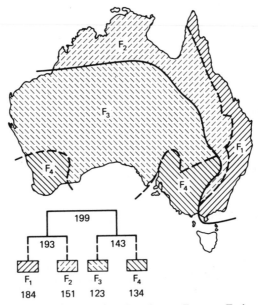

Figure 1.4 Division of Australia into four faunal areas $F_1, \ldots, F_4$, based on the distribution of 199 genera of land birds at 121 sites. The sites, scattered over the continent, are not shown. The dendrogram on which the classification is based is shown below the map. The numbers are the numbers of genera in the sites undergoing subdivision at each stage. (Adapted from Figure 3 of Kikkawa and Pearse, 1969, who continued the subdivision until 10 areas had been differentiated.)

ranges of individual species form line segments.* If a large number of these line segments are coincident (or approximately so), one can reasonably regard the corresponding stretch of shore as a biogeographic unit. In theory, the boundaries of such a unit are merely two points. In practice, of course, the boundaries may be rather fuzzy, since the range extremities of different species seldom coincide exactly.

To classify a stretch of shoreline in this way, it is necessary to marshall the relevant data on shore organisms' ranges in an easily interpretable form. Two ways of doing this are shown in Figure 1.5.

For ease of exposition it is assumed that the shoreline concerned trends north-south and for the moment only two species, A and B, are considered. Their ranges are shown in Figure 1.5a. Species A's range is from 50°N to 20°N and species B's from 30°N to 40°S. Each species' range is representable by a point in a two-dimensional graph and two convenient ways of constructing the graph's axes are shown in Figure 1.5b, c.

Figure 1.5b shows Pielou's (1977b, 1978b) method: the coordinates are the midlatitude of a species' range (on the abscissa) and the magnitude, or *span*, of its range (on the ordinate). The edges of the triangle bound the space within which all points must lie. This is because it is assumed that the extremities of all species' ranges are contained within the map area; therefore a species whose midlatitude is close to one end of the shoreline must necessarily have a small span.

Figure 1.5c shows Hayden and Dolan's (1976) method of graphing ranges. The coordinates are the southern and northern latitudinal limits of a species' range. Again, all possible points are constrained to lie within a triangle; this simply arises from the fact that the northern extremity of a range cannot be south of its southern extremity.

The coordinates of a given point graphed by the two methods are related as follows.

Denote the midlatitude and the span of the species' range by x_1 and y_1 respectively; these are the coordinates of the points in Pielou's graph.

Denote the southern and northern end points of the range by x_2 and y_2 respectively; these are the coordinates of the point in Hayden and Dolan's graph.

Note that x_1, x_2, and y_2 are negative if they represent southern latitudes. Then it is easy to see that

$$x_1 = \tfrac{1}{2}(x_2 + y_2), \qquad y_1 = y_2 - x_2;$$

*Note that this discussion is about the ranges of shore organisms on a geographic, not an ecological, scale. We are here *not* concerned with the small-scale ecological pattern resulting from zonation parallel with the sea's edge.

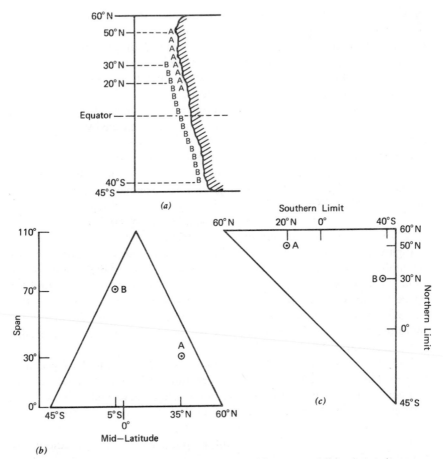

Figure 1.5 Two ways of representing the geographic ranges of littoral organisms on a north-south trending shore. (*a*) An imaginary shoreline extending from 60°N to 45°S and the ranges of two littoral species, A and B. (*b*) Pielou's (1977b, 1978b) method of representation. (*c*) Hayden and Dolan's (1976) method of representation.

or, equivalently,

$$x_2 = \tfrac{1}{2}(2\,x_1 - y_1), \qquad y_2 = \tfrac{1}{2}(2\,x_1 + y_1).$$

Hayden and Dolan (1976) plotted (by their method) the ranges of 968 littoral species occurring on the Atlantic and Pacific coasts of the Americas; the organisms considered were chiefly ascidians, crabs, and mollusks. When several species are found to have approximately coinci-

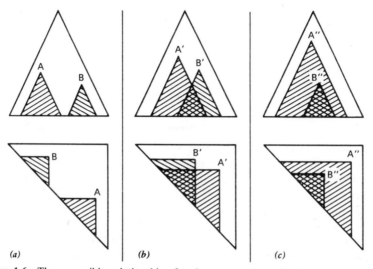

Figure 1.6 Three possible relationships for the ranges of a pair of littoral species (or species-groups), each plotted in two ways. The ranges (a) of species A and B are disjunct; (b) of species A' and B' are overlapped; (c) of species A" and B" are nested.

dent ranges, or equivalently, when the points representing these species form a compact cluster, the joint range of the whole group of species is called by Hayden and Dolan the group's *co-range*. Hayden and Dolan further define such co-ranges as "coastal-marine-biotic provinces."

It should be noticed, however, that provinces so defined can overlap each other, or the smaller can be nested within the larger. To see this, consider Figure 1.6. The axes of the triangular graphs are as in the similar graphs in Figure 1.5. The same two points are plotted in each vertical pair of graphs. Further, each point has been made the apex of a shaded triangle whose sides are parallel with those of the outline triangle of the graph containing it. It will be seen that, depending on whether the geographic ranges of a pair of species are disjunct, overlapped, or nested, the shaded triangles associated with their points are likewise disjunct, overlapped, or nested.

Now suppose that the points A' and B', for example, are each replaced by a compact cluster of points, with one point representing one species. It follows that the species in each cluster have a common co-range. The relative arrangement of the two co-ranges is as indicated by the shaded triangles; thus, in the case now being considered, they overlap. Since biogeographic units are customarily defined so as not to overlap, it would lead to confusion to describe a co-range as a "province." The word *co-range* can usefully serve as it stands, as the name for units that are permitted to overlap.

Use of a Similarity Matrix

In partitioning a shoreline into biogeographic units, one would like to be sure that each unit is internally homogeneous and, at the same time, that the different units differ distinctly from each other. Such a partitioning may be unattainable. If the change in the flora and fauna along the shore is gradual and steady, with no discontinuities, then any partitioning is necessarily artificial. Therefore it is useful to be able to test (informally) whether natural discontinuities, that could form the basis of a classification, do in fact exist.

One way to judge is to compile a *similarity matrix*. The data needed are lists of the species occurring within small, equal-sized sampling intervals along the shoreline. For example, Valentine (1966), who partitioned the Pacific shore of North America on the basis of the benthic mollusk fauna, used single degrees of latitude as sampling intervals.

One then calculates the "similarity" of all possible pairs of intervals and plots the result as a matrix (see Figure 1.7). There are numerous ways in which similarity can be measured; they have been reviewed in detail by Cormack (1971) and Goodall (1973). Two frequently used coefficients are Jaccard's (used by Valentine) and Sorensen's. They are defined as follows.

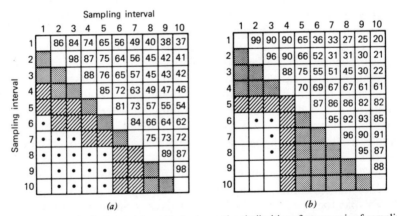

Figure 1.7 Two similarity matrices. Each shows the similarities of every pair of sampling intervals in a row of 10 such intervals. The numbers in the cells above the diagonals are percentage similarities. In the cells below the diagonals, these numbers, instead of being repeated, are replaced by shading. The shadings correspond to similarities of >80, 60 to 80, 40 to 60, and <40.

Denote by a the number of species common to both sampling intervals, and by b and c respectively the number of species occurring in only one of the two intervals. Then Jaccard's coefficient is $a/(a + b + c)$ and Sorensen's coefficient is $2a/(2a + b + c)$.

The matrix of similarity coefficients is, of course, symmetrical. If the cells in one half of it are shaded as shown in the figure, with dark shading for high similarity and light shading for low, one obtains an immediate visual impression of the way the different sampling intervals are related. Thus consider the two contrasted examples shown in the figure (for clarity, the number of sampling intervals has been made unusually small). It is obvious that the "shore" in Figure 1.7a exhibits gradual, continuous variation, whereas that in Figure 1.7b is clearly separable into two very distinct segments.

One can therefore judge, from the appearance of the similarity matrix, whether a natural classification of the data is possible or not. Whatever the outcome, a classification can always be done, by any one of a vast number of different "clustering algorithms" [Gnanadesikan, Kettering, and Landwehr (1978) and also Frenkel and Harrison (1974) discuss the problem of choosing among the myriad possibilities]. In any case, although the methods can be counted upon to classify any body of data, there is no guarantee that the classes so recognized are naturally distinct. They may be entirely artificial. The purpose of inspecting a similarity matrix is to decide this point. If the matrix chances to yield very distinct and clear-cut classes (rather an unlikely event), one may simply use the classification suggested by the matrix as it stands, without further computation. This is because no one clustering method is intrinsically better than any other.

It should also be emphasized that one cannot use a single body of data both to arrive at a classification and also to test, statistically, whether the classes differ from each other significantly. If testing is desired, it must be performed on a separate body of data, independent of that on which the classification is based. These remarks apply to all classifications, of course, not only to the partitioning of a linear row of samples.

Chapter Two

BIOGEOGRAPHY AND CONTINENTAL DRIFT

During the 1960's and 1970's a revolution has been taking place in the earth sciences. The newly developed theory of plate tectonics, which is now supported by overwhelming evidence, has provided a mechanism to account for "continental drift." The number of earth scientists who are convinced that the continents have drifted, and are still drifting, over the earth's surface, has swelled from an eccentric few to the great majority. Simultaneously, great changes have taken place in biogeography. While some biogeographers have always assumed the reality of continental drift despite the derision (in the past) of most geophysicists, others, assuming that geophysics had attained a final form which absolutely precluded continental drift, busied themselves devising improbable feats of dispersal or geophysically unlikely land bridges to explain the distribution of plants and animals. All that is now over. Biogeographers and geophysicists now see eye to eye and the interaction between them is leading to rapid advances.

This chapter is concerned with some of these advances. Because of the vastness of the subject, it has been necessary to be very selective in choosing topics to include. These have been grouped into four sections.

The first discusses paleogeography, the way the configuration of the earth's land masses has varied through geological time, particularly during the last 230 million years. This provides the background for the steadily changing distributions of animals and plants. Animals and plants are themselves changing, of course, that is, they are continuously evolving, as their geographic distributions change. Consequently the organisms themselves, as well as their geographic patterns, are very different now from what they were in the geological past.

Sections 2, 3, and 4 consider respectively plants, vertebrates, and invertebrates. We are concerned with long-term changes in the biogeography of these groups, and with the way in which continental drift has brought these changes about.

1. PLATE TECTONICS AND PALEOGEOGRAPHY

According to modern plate tectonics theory, the lithosphere, that is, the rigid outer shell of the earth, is made up of a number of separate plates. Opinion as to the number of them differ; there may be as many as 10 (Dietz and Holden, 1970) or as few as 6 (Le Pichon, 1968). Their thickness is probably about 100 km. The plates rest on a more malleable underlying layer, the asthenosphere, and move relative to each other upon this layer.

The movements of the plates take various forms. They may separate and drift apart as the sea floor between them spreads along midocean fissures. They may converge and collide with each other; when this happens, one of the colliding plates rides up and the other dips beneath it, along a so-called subduction zone. They may shear past each other. Or they may rotate.

Each plate is a rigid segment of the earth's shell and is constructed of one or (in places) two layers. The lower layer consists of dense rock (chiefly basalt), known as sima because it is rich in silica and magnesium. This is the rock that forms the earth's crust at the bottom of the ocean basins. Plates having only this one layer are wholly submerged.

Other plates have two layers. On top of the lower layer of sima are segments of an upper layer of comparatively light rock (chiefly granite), known as sial because it is rich in silica and aluminum. This is the rock that forms the continental part of the earth's crust. This lighter, continental crust occurs as superimposed attached "blocks" (that is, continents) upon the heavier, oceanic crust. The continents are smaller in extent than the plates themselves and a continent drifts as it is transported, passively, on the "back" of the plate on which it lies. That is, as the plates move they carry the continents with them.

The foregoing brief account is based on numerous sources, chief among them Wilson (1963a). The units whose movements geophysicists are concerned with are the plates themselves, regardless of the fact that some parts of a plate's upper surface (the part consisting of oceanic crust or sima) are submerged beneath the sea and other parts (the part consisting of continental crust or sial), if present, are exposed as dry land. For the biogeographer, however, the distinction between land and sea is fundamental. It is the drift of the continents rather than of the plates per se that governs the distributions of terrestrial plants and animals. Therefore the older term "continental drift" more accurately describes a biogeographer's preoccupation than does the newer term "plate tectonics."

The speed at which drift takes place varies. Le Pichon (1968) reports speeds of between 1 and 10 cm per year.

We now consider the history of the earth's land surfaces over the past 230 million years, that is, since the end of the Permian, the last period of the Paleozoic era. At that time, it is believed, all the continents were joined to form one single continent, "Pangaea," surrounded by a single universal ocean, "Panthalassa." The area of Pangaea was approximately 200 million km^2, or 40% of the earth's total surface; this is roughly the same as the present area of the continents, including their shelves, as bounded by the 2000 m depth contour.

Then, about 200 m.y. ago, at the beginning of the Mesozoic era, Pangaea began to split into fragments. The events that have occurred since that time are, of course, extraordinarily complex and very incompletely known. As a basis for further discussion, it is helpful to consider maps (Figures 2.1 and 2.2) and a table (Table 2.1) that summarize as compactly as possible some widely held views on the subject. They combine, and as far as possible reconcile, the synopses of Dietz and Holden (1970), Keast (1973), and Raven and Axelrod (1974).

First consider Figure 2.1. The areas outlined as continents on geophysicists' maps are "whole" continents; that is, they coincide with the pieces of continental crust (composed of sial) that form the upper, discontinuous layer of the tectonic plates. Not separately distinguished, usually, are those parts of the continents temporarily submerged to form epicontinental or shelf seas. These seas are not permanent. The land underlying them is alternately submerged and exposed as the sea level changes. To repeat: they are part of the continents, not of the deep sea which is everywhere underlain by oceanic crust (composed of sima), the permanently submerged lower layer of the tectonic plates.

However, to a terrestrial organism, sea constitutes a barrier to dispersal whether it be shallow or deep; and to a benthic marine organism, shallow shelf seas (or epicontinental seas) provide a home and dry land does not. Therefore the changing pattern of land and sea is even more complicated to a biogeographer than to a geophysicist. The latter has only two entities to consider: deep sea and "whole" continents. For the biogeographer there are three: dry land, shallow sea, and deep sea. For this reason, epicontinental seas have been sketched in on the maps of Figure 2.1. The purpose is more to keep the reader constantly reminded of their existence

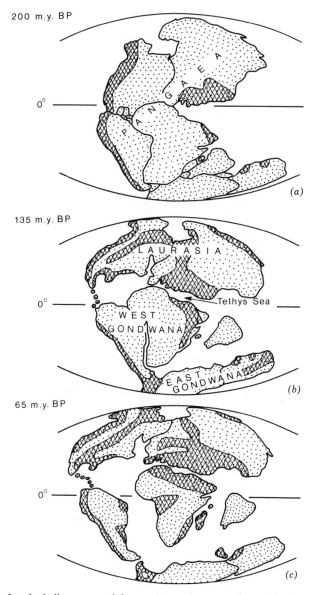

Figure 2.1 Land, shallow sea, and deep ocean at three past times. The hatched areas are epicontinental seas. Plate boundaries are not shown. (*a*) Pangaea in the Triassic. (*b*) Late Jurassic or early Cretaceous. Northern Pangaea has become Laurasia; southern Pangaea, after becoming Gondwana, has split again into West Gondwana, East Gondwana, and India. (*c*) Late Cretaceous. (Adapted from Holden and Dietz, 1971, Kummel, 1970, and Tedford, 1974.)

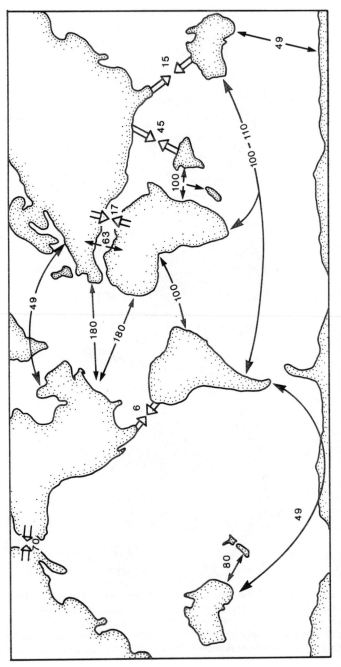

Figure 2.2 Estimates of the times (in millions of years BP) at which direct dispersal routes between land masses were made or broken. Hollow converging arrows show joins (the "join" between Australia and Asia refers to the narrowing of a gap and the appearance of stepping stone islands). Black diverging arrows show separations. (Based on Raven and Axelrod, 1974.)

Table 2.1 Estimated times of some of the chief events in earth history since the early Mesozoic. Double-headed arrows denote separations. Names of paleocontinents are in italics. (Based on Dietz and Holden, 1970; Keast, 1973; Raven and Axelrod, 1974; and McKenna, 1975.)

Period or epoch	Time (m.y. BP)	Event
Early Triassic	⩾200	The continental crust formed a single continent, *Pangaea*.
Late Triassic	180	*West Laurasia* (North America) ⟷ Africa. *West Gondwana* (Africa + South America) ⟷ India ⟷ *East Gondwana* (Australia + Antarctica).
Early Cretaceous	135–125	South America ⟷ Africa in far south because of rotational movement.
Mid-Cretaceous	110–100	South America ⟷ Africa at latitude of Brazil. Africa ⟷ Madagascar ⟷ India. Africa, India, and Australia all drifting northward.
Late Cretaceous	80	North America ⟷ (Europe + Greenland). (Antarctica + Australia) ⟷ (New Zealand + New Caledonia).
Very late Cretaceous	70	Contact made between northwestern North America and northeastern Siberia.
Very early Paleocene	63	Africa ⟷ Europe (temporarily).
Eocene	49	Dispersal route between North America and Eurasia, from being predominantly via North Atlantic, switches to Beringia because North Atlantic becomes wider and Beringia becomes warmer.
Eocene	c49	Australia ⟷ Antarctica.
Eocene	45	India drifts into contact with Asia.
Oligocene	c30	Turgai Strait (east of Ural Mountains) finally dries up.
Miocene	17	Europe and Africa rejoined.
Miocene	15	The narrowing gap between Australia and Southeast Asia, and the appearance of stepping stone islands permits plant dispersal.
Pliocene	6	North America and South America joined by land bridge.

than to show their true outlines at any time. Comparatively minor warp-
ings of the earth's crust, with accompanying simultaneous changes in both
the shape of low-lying land areas, and in the depth of the sea, can
markedly affect the extent and shape of epicontinental seas. These
changes are usually more rapid than those brought about by the slow
movements of the plates.

Even so, faunal separations caused by these seas are detectable and
permit the recognition of paleobiogeographic regions on the basis of fossil
evidence. For example, according to Cox (1974) there were two distinct
dinosaur faunas in the late Cretaceous in what are now the northern
continents. One occupied "Asiamerica," that is, western North America
linked to Asia across an unsubmerged Bering Strait ("Beringia"). The
other occupied "Euramerica," that is, eastern North America linked to
Europe across the then closed North Atlantic. Separating these faunas
were two epicontinental seas: the Midcontinental Seaway of North
America, which extended the whole length of the continent east of the
present Rocky Mountains; and the Turgai Strait, which likewise formed a
north-south barrier between the present Europe and Asia and finally
disappeared in the Oligocene (McKenna, 1975).

Next consider Figure 2.2. It is an attempt to present the material in
Table 2.1 in pictorial form. The arrows in the figure show "separations"
and "joins." Two land masses become biogeographically separated when
they drift apart or when what had been an easy dispersal route between
them becomes impassable owing to climatic change. An apparent overland route that is too cold or too arid is not a route at all. Likewise, two
areas become biogeographically joined when they drift into contact with
each other or when the climate of a previously unusable land bridge
improves.

A third possible cause of "joins" is the appearance of stepping stone
islands forming a dispersal route across a barrier sea. Such a route is not
passable for all organisms; what is an easy route for wind-dispersed plants
and many insects may be completely impassable for mammals.* Stepping
stone islands can appear either because of volcanic activity or because of
crustal warping. For example, volcanic islands provided a stepping stone
route between North and South America before the Isthmus of Panama
came into existence (Malfait and Dinkelman, 1972) (see Figure 2.3). Thus
the ease of dispersal between the two continents has varied, being harder
at some times than at others.

The stepping stone islands linking the Oriental and Australasian regions

---

*Chapter 8 contains a fuller discussion of dispersal of various kinds of organisms, by various
routes.

(that is, the Indonesian islands) were formed by crustal warping. As a dispersal route, they are still impassable to nearly all mammals (see Figure 1.2).

If the appearance of stepping stone islands in a water gap can create a biogeographic join then, obviously, their disappearance can bring about a biogeographic separation. No examples of this phenomenon appear to be known, however.

It must now be emphasized that many of the assertions embodied in Figures 2.1 and 2.2 and Table 2.1 are unacceptable to many authorities. It seems safe to say that there is general agreement about the broad outlines of continental movement. Few people now doubt that sometime in the Mesozoic all the world's land constituted two great continents, Laurasia and Gondwana.* Laurasia was made up of what are now North America and Eurasia, Gondwana of what are now Antarctica, Australia, South America, Africa, and India.

All other points are more or less contentious. As examples it is worth listing, in numbered paragraphs, six of the more important controversies, some of which have great biogeographic significance. They are as follows:

1. Although early Mesozoic terrestrial faunas appear to have been cosmopolitan (Charig, 1973), not all workers are convinced that Laurasia and Gondwana were ever joined to form the single continent of Pangaea.

2. There is considerable doubt as to whether, before Gondwana broke up, the east coast of India was contiguous with Antarctica (as in Figure 2.1a) or with Australia.

3. From the biogeographic point of view, the history of the Tethys Sea is most uncertain. At the beginning of the Mesozoic it formed a deep gulf in the east coast of Pangaea (if there was a Pangaea) or was the eastern half of a continuous channel separating Laurasia and Gondwana (if there was no Pangaea). In either case, its subsequent history was largely determined by rotational movements, clockwise in the case of Laurasia and counterclockwise in the case of West Gondwana (see Figure 2.1). The coming together, like a pair of jaws, of the shores of what are now North Africa and southwest Eurasia led to crustal downwarping (that is, the development of a geosyncline) and hence to the formation or widening of epicontinental seas. These very complicated events make it hard to judge when and where dispersal routes formed, and how long they persisted (Raven, 1979).

*The southern continent is more often called *Gondwanaland* than *Gondwana*. Nothing seems to be gained by sacrificing brevity and it is here called *Gondwana*.

4. The history of the Central American Isthmus is much more complicated than Figure 2.2 implies. Although the current land bridge, which was formed about 6 m.y. ago, is probably the first unbroken land link between the Americas, there may have been earlier, stepping stone routes across the gap. According to Malfait and Dinkelman (1972; and see Rosen, 1975), in the late Jurassic and early Cretaceous, before West Gondwana split, its western half (South America) was linked to the western half of Laurasia (North America) by a chain of islands (see Figure 2.3a). As the continents drifted westward, the tongue of crustal plate between them, on which the islands were located, moved relatively eastward. They are now the Greater Antilles. Meanwhile, a new generation of volcanic islands, the Central American Archipelago, appeared in the intercontinental gap and ultimately fused to form the present isthmus. It is difficult to infer from this history whether, and if so when, a dispersal route passable by land mammals existed.

5. A "join" that few authors discuss is that between the present Alaska and Siberia. As North America and Eurasia drifted away from each other on either side of the widening Atlantic, their farther shores approached each other across the narrowing North Pacific. The Dietz and Holden reconstruction, as shown in Figure 2.1, makes it appear that the line of collision was in what is now Bering Strait. Wilson's (1963) reconstruction, according to which the Verkhoyansk Mountains of northeastern Siberia mark the line of collision, seems more likely. It implies that the region known as Beringia, consisting of the easternmost tip of Siberia, Alaska, and the continental shelf between and around them, formed the westernmost part of Laurasia (see Figure 2.4). The uplift of the Verkhoyansk Mountains, which was caused by the collision of the plates, began in the early Cretaceous about 120 m.y. ago. Formation of a continuous land surface would have come much later, according to McKenna (1975) about 70 m.y. ago.

6. The northern border of the Indian fragment of Gondwana has been thought to coincide with the line of the present Himalayas. However, Crawford (1974) considers that Gondwana was considerably bigger and extended for thousands of kilometers beyond this line. That is, much of what is now western and central China formed part of Gondwana and the Himalayas are an *intra*continental range and do not, as had been thought, mark the edge of a crustal plate. Biological evidence in support of this view comes from two very different organisms: the Cladoceran genus *Daphniopsis* which is found in Kerguelen, Antarctica, Australia, Tibet, and Inner Mongolia; and the fossil tetrapod *Lystrosaurus*, known from Antarctica, South Africa, India, and Sinkiang and Shansi in China (see Section 2.3). These ranges imply either that Crawford's surmise is cor-

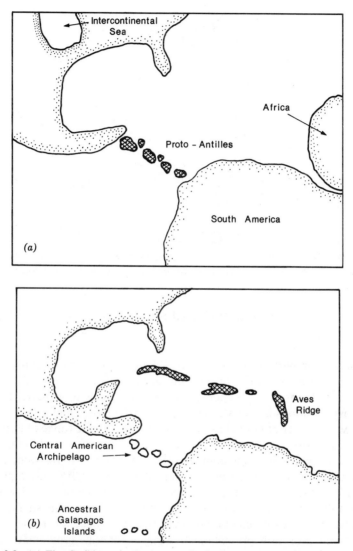

Figure 2.3 (*a*) The Caribbean in the very early Cretaceous (contemporary with Figure 2.1*b*). The South Atlantic has not yet opened up and an epicontinental sea still covers much of North America. The Americas are linked by the Proto-Antilles (hatched) which will be forced eastward, relatively, as the continents drift west. (*b*) Late Tertiary. The Proto-Antilles have become the Greater Antilles. The Lesser Antilles will appear as volcanic islands on the slopes of Aves Ridge. New volcanic islands, the Ancestral Galapagos, and the Central American Archipelago have appeared. The latter will consolidate to form the Isthmus of Panama. (Adapted from Malfait and Dinkelman, 1972, and Rosen, 1975.)

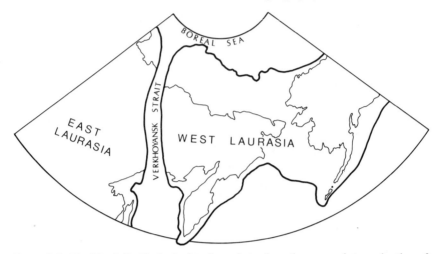

Figure 2.4 The North Pacific in the late Jurassic to show the approach to each other of East Laurasia (Siberia) and West Laurasia (North America). (Adapted from Wilson, 1963a.)

rect, or that the organisms dispersed before Pangaea broke up; perhaps both statements are true.

A final very important point to notice, before we come to consider the effect of continental drift on specific groups of organisms, is that the drifting continents have experienced, and continue to experience, tremendous variations in climate. This results from three chief causes. In the first place, most continents have drifted through a wide range of latitudes. Figure 2.5 shows, as examples, how the positions of three familiar land masses have changed in the course of the last 200 m.y. In the same period the southern tip of India has moved from about 60°S to 8°N, and Japan has moved from the Arctic to its present position in warm temperate latitudes.

Secondly, the splitting of Laurasia and Gondwana into smaller segments has brought the sea close to areas that were once far inland. Moist, maritime climates now prevail in some places that had previously experienced great aridity and the temperature extremes of continental interiors.

Thirdly, the upwarping that accompanies drift in some places has converted lowlands into highlands. The climates of these lands have changed correspondingly and, in addition, rain shadows have been created where there were none before.

Besides causing great change in the temperature and moisture regimes of an area, the rearrangement of the continents has also brought changes in the strength and direction of prevailing winds with important consequences for wind-dispersed organisms.

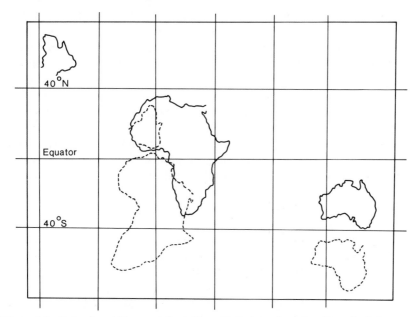

Figure 2.5 Labrador, Africa, and Australia, with their modern shorelines, in their present positions (solid outlines) and their positions in the Triassic, 200 m.y. BP (dashed). (Adapted from Dietz and Holden, 1970.)

Thus in reading the following sections it should be kept in mind that the dispersal of terrestrial organisms depends on more than the lie of the land, that is, on the existence of land bridges or stepping stone routes. To be usable, potential dispersal routes must have climates that the dispersing organisms can endure. And the migrations of wind-dispersed organisms are strongly affected by prevailing winds. The geographic distributions of animal species are also dependent, of course, on the distribution of vegetation, that is, of plants. In discussing the effects of continental drift on plant biogeography in the next section, therefore, it is important to consider the geography of both taxonomic and ecological units.

2. PLANT DISTRIBUTION AND CONTINENTAL DRIFT

From the breakup of Pangaea near the beginning of the Mesozoic era, to the end of the Tertiary period* about 2 m.y. ago, covers a time span of

---

*Events in the last two million years, that is in the Quaternary period or ice age, are treated in chapter 4.

perhaps 200 m.y. During that time the continents have drifted, and their climates have changed both because of the drift and because of worldwide climatic variation. While this has been going on, biological species have been migrating and have simultaneously been evolving. Because of the continuous change due to evolution, we cannot assume that a species' present-day ecological requirements match those of its distant ancestors. But the ecology of a species known only as a fossil can at any rate be inferred from its anatomy and morphology. In any group of organisms there is probably a fairly close correspondence between the rates of evolution of their morphological and of their ecological characteristics [for a contrasting opinion, see Kornas (1972)].

Evolution in the Angiosperms

In attempting to trace how the geographic patterns of modern plants have developed since the early Mesozoic, we must begin with a discussion of evolutionary rates, and the times at which different groups have come into existence so far as is known from the fossil record. We shall be chiefly concerned with the evolutionarily most advanced plants, the angiosperms, or flowering plants. Their origin continues to baffle all who attempt to unravel it. Thus Beck (1976) wrote, ". . . the mystery of the origin and early evolution of the angiosperms is as pervasive and as fascinating today as it was when Darwin emphasized the problem in 1879." There is no consensus on when they originated, where they originated, or how they originated, or even on whether they are monophyletic (the majority opinion) or polyphyletic.

The question of "when," in particular, is a topic of much disagreement. It hinges upon whether fossils of vegetative plant parts such as leaves, dating from the Triassic and Jurassic, came from true angiosperms or from angiosperm precursors. The oldest fossils that are unquestionably angiosperms, since they bear angiospermous fruits, date from the early Cretaceous. Beck (1976) is of the opinion that this date marks the origin of the angiosperms. Axelrod (1970, 1972) puts their origin much earlier, at the end of the Triassic or beginning of the Jurassic. Treating these two dates as extremes, the range of possibilities thus covers a span of about 50 m.y. (that is, from 180 m.y. BP to 130 m.y. BP). This is more than 20 times as long as the interval between the beginning of the Quaternary ice age and the present. It provides enough time for the continents to shift their positions relative to each other by as much as 2500 km (about the distance between Africa and South America now) assuming an average drift speed of 5 cm per year.

Clearly, no history of the spread of the angiosperms can be regarded as trustworthy until the mystery of their origin is solved. However, to provide a base for further discussion, it is desirable to have in mind one possible outline, that of Axelrod (1970, 1972). His reconstruction is briefly as follows. He assumes that the angiosperms originated in the Triassic-Jurassic transition, in tropical uplands with a warm, equable, seasonally dry climate, probably in the interior of West Gondwana (Raven and Axelrod, 1974). The property of angiospermy, that is, the possession by a plant of enclosed ovules, probably evolved where it was selectively favored, in regions of seasonal drought. Therefore, if this is true, the earliest angiosperms were not at sites such as low-lying swamps where their remains would stand a good chance of being preserved as fossils. Thereafter, the newly evolved angiosperms expanded their ecological tolerances, and hence their geographic ranges, in the manner shown schematically in Figure 2.6. Progressively, over several million years, they invaded regions where temperatures were less equable and more extreme, either hotter or colder; altitudinally, they migrated downward into hot lowland and upward into cool mountains; latitudinally, they migrated northward and southward from the equator.

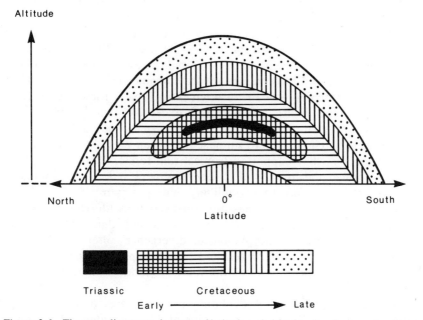

Figure 2.6 The spreading range, in terms of latitude and altitude, of angiosperm dominance from their original appearance, probably in the late Triassic, at medium altitudes in tropical West Gondwana. (Adapted from Axelrod, 1970.)

In support of this thesis, Axelrod (1970, 1972) notes that, in every class of angiosperm, the more primitive families and genera are now found in warm temperate or high-altitude tropical regions, whereas the more advanced taxa occur predominantly in less equable regions. Similarly, Takhtajan (1969) notes that forests of the subtropics and, within the tropics, those high on mountains, contain more representatives of primitive families than do tropical lowlands.

To assume, as Takhtajan and Axelrod do, that the most primitive taxa are found in, and hence are indicative of, a group's ancestral region is to take one side in a debate that is far from settled. It is discussed in some detail in Chapter 3.

However, regardless of whether advanced taxa are to be expected at the center or the periphery of an expanding range, evolutionary advances are always being made somewhere and it is necessary to consider evolutionary rates. In reconstructing the simultaneous progress of biological evolution on the one hand, and continental drift on the other, one must ensure that the postulated order of events is reasonable. For example, the presence in both Australia and New Zealand of a family known to have been in existence for 100 m.y., say, is consistent with the beliefs that Australia and New Zealand separated 70 m.y. ago and that no member of the family could have dispersed across the present Tasman Sea.

But it is not reasonable to adduce as evidence for continental drift the supposed presence in both tropical West Africa and Brazil of the same species of rain forest tree, *Symphonia globulifera,* and at the same time to suppose that the two land masses separated somewhere between 70 and 100 m.y. ago; (the figure of 100 m.y. given in Table 2.1 and Figure 2.2 is probably an upper limit). Not only is it excessively improbable, as we shall see below, that a taxon of the level of species should persist for so long, there is also no reason to suppose that when West Africa and Brazil were contiguous they supported tropical rain forest. At that time, they together formed the interior of a supercontinent and were presumably much drier. It therefore seems unlikely that specimens of *"Symphonia globulifera"* from the two areas are conspecific, as Melville (1973) has asserted. Their resemblance probably shows only that they have not undergone conspicuous morphological divergence; alternatively, their geographic range could have resulted from comparatively recent long-distance dispersal, perhaps when the south Atlantic was narrower than it is now.

The rates at which plant evolution proceeds are, of course, very difficult to infer. Different groups evolve at different rates, and within a group the rate is unlikely to remain constant over long stretches of time. The easiest way to make visualizable what is known (or supposed) is to

list what proportions (in very general terms) of modern plant taxa of
different taxonomic levels are believed to have been in existence in
successive geological periods and epochs. Summarizing in this form the
opinions of Raven and Axelrod (1974), who discuss the early history of
the angiosperms in great detail, gives:

Mid-Cretaceous.   In pollen-bearing sediments, for the first time, the
proportion of angiosperm pollen exceeds the proportion of gymnosperm
pollen and fern spores combined. Among angiosperms, many modern
orders and some families are in existence, but no modern genera.

Late Cretaceous.   Some modern families have evolved, such as the
Proteaceae (whose extant members include the sugar bushes of South
Africa and the silky oaks of Australia), the Graminae (the grass family)
and the Myrtaceae (the eucalypt family); and some modern genera, such
as *Alnus* (alder) and *Ilex* (holly).

Paleocene.   Most modern families exist, and many more modern genera,
such as *Betula* (birch) and *Liquidambar* (sweet gum).

Lastly, Leopold (1967) states that in the:

Pliocene.   Nearly all angiosperm genera and the majority of species are
in existence.

It is interesting, though difficult, to compare evolutionary rates in
entirely different classes of organisms. It is a tautology to say that primitive organisms evolve more slowly than advanced ones. This must be
true: an advanced taxon must have evolved fast to have reached its level
of advancement. Knowledge about relative rates of evolution is certainly
worth striving for, however.

According to Leopold (1967), most living species of conifer were in
existence in the Miocene but probably fewer living species of angiosperm
date back quite so far. Although angiosperms are the most rapidly evolving plants, they evolve much more slowly than mammals, and probably
rather faster than mollusks. Thus one need go back only to the mid-
Pleistocene to reach a time when the majority of modern mammal species
had not yet come into existence, whereas the time at which the same was
true of mollusk species was the early Miocene.

To make more definite, numerical statements about relative evolution
rates seems unwarranted, for two reasons. In the first place, all that is
known is based on the evidence of fossils and therefore depends on which
plants chanced to become fossilized, and which fossils chanced to be
discovered and identified. The second, more fundamental difficulty is that
the inclusiveness of the entity "species" varies from group to group.

Stegman (1963) considers that a botanical species is more nearly equivalent to a zoological subspecies than to a zoological species, because plant taxonomists are more prone to "splitting" than are animal taxonomists. Hence, if calculated rates of evolution are based on estimates of the turnover rates of named species, the rates for plants will be overestimated relative to those for animals. If this is so, the effect is to reinforce the conclusion that angiosperms evolve much more slowly than mammals.

An evolutionary advance in plants that is very important from the biogeographic point of view is the continual improvement, over time, of their dispersal capacities. According to Schuster (1976), long-distance dispersal mechanisms are chiefly of Tertiary origin. Mechanisms such as edible fruits, that depend for their success on the presence of birds and mammals to transport them, have evolved simultaneously with fruit-eating birds and mammals. Mechanisms that enable seeds to be dispersed by wind evolved also. Primitive plants had heavy, wingless seeds. As examples of angiosperm families within which primitive and advanced dispersal mechanisms may still be found, Schuster mentions the Ranunculaceae and the Rosaceae. The fruits of some well-known genera in these families are as follows:

	RANUNCULACEAE	ROSACEAE
Primitive:	*Aquilegia* (dry follicle)	*Spiraea* (dry follicle)
Advanced:	*Actaea* (fleshy berry) *Clematis* (persistent style = "parachute")	*Fragaria* *Rubus* *Rosa* *Prunus* *Malus* (fleshy fruits)

Changes in Climate and Vegetation

As the continents drifted and plants (and animals) evolved, climate, and hence vegetation, were undergoing continuous change. Opinions on climatic change are based on fossil finds that are scattered both geographically and with respect to time. Thus even if the climate prevailing at a particular time and place has been correctly inferred, there is often no way of knowing how long the conditions persisted, or over how extensive an area.

The following is an attempt to collate and reconcile the views of several authors on how climates have changed since Mesozoic times until the end of the Tertiary period 2 m.y. ago.

It seems to be generally held that the Cretaceous was the warmest period in the last 200 m.y. Axelrod (1970, 1972) considers that it was, as well, the wettest period. Epicontinental seas of great extent existed; the opening of the Atlantic caused the opposing coasts of West Africa and South America to have wetter climates than they had when they were interior parts of a single West Gondwana. Generally, wetter climates had the effect of expanding the ranges of moist forest species and restricting the ranges of many xerophytic plants that had evolved earlier. The temperature gradient from equator to poles was less steep than it is now. Trees grew in Antarctica (Llano, 1965), which owed its favorable conditions both to the worldwide warmth of the climate and to the fact that it had not yet drifted into the south polar region.

From the Cretaceous onward into the middle and late Tertiary, the general tendency was everywhere toward cooler, drier climates and a more pronounced latitudinal gradient, but over so enormous a time span it seems certain that the general trend was subject to many temporary reversals. In any case (Axelrod, 1970, 1972), as Antarctica drifted southward the vegetation appears to have changed from forest to fern scrub and then to cushion plant communities as the continent gradually cooled. The Midcontinental Seaway of North America dried up. But Africa remained much wetter than it is now and the whole continent, including the present Sahara Desert, was forested. Then, upwarping of the continental crust raised the center of Africa to altitudes of perhaps 3000 m. The newly formed uplands were cooler, because of altitude, and in many regions drier, because of the deflection of rain-bearing winds, than they had been previously. Many rain forest species became wholly extinct; others became disjunct because of extinction in the uplands and now no longer occur in central Africa but only in the extreme east and west; an example is shown in Figure 9.3. At the same time, however, volcanism and the formation of the East African rift valleys with their bordering highlands greatly diversified the topography; some cool to cold high-altitude regions with ample rainfall came into existence and provided environments for temperate montane rain forest and alpine vegetation (Axelrod and Raven, 1978).

Events in the Antarctic are of particular interest since they have much more than local importance. Thus Raven and Axelrod (1974) have argued that formation of an Antarctic ice sheet, and the lie of the southern continents, resulted in the flow of cold ocean currents. Since the appearance of the ice sheet, cold Antarctic waters have been brought to the western shores of South Africa by the Benguela current, and to the western shores of South America by the Humboldt current. As a result, climates on these shores have become colder and drier and there has been a sequence of vegetation changes; following the earlier rain forests have come, in turn, savannah, thorn scrub, and now desert.

If ocean currents cooled by Antarctic ice have caused these changes, their times of occurrence obviously depend on when the ice sheet formed. This is a topic of considerable disagreement. For example, Raven and Axelrod (1974), on the basis of botanical evidence, believe the glaciation of Antarctica to have begun at least 20 m.y. ago in the mid-Miocene; the ice sheets then expanded, and had covered the whole continent by 4 to 5 m.y. ago, long before the end of the Pliocene. Margolis and Kennett (1971) arrive at rather different conclusions, using as evidence the fossil Foraminifera in deep-sea cores from the Antarctic Ocean. They consider that Antarctica was glaciated to some extent in the lower and middle Eocene and the Oligocene, but not necessarily continuously. Ice sheets of various sizes may have formed and melted repeatedly. They believe that the warmest part of the Cenozoic in Antarctica was the early and middle Miocene when there may have been no ice sheets, and that continuous cold and heavy glaciation did not begin until the late Miocene.

Wolfe (1971) deduces past climatic conditions not from the taxonomic identity of fossils, but from their physiognomic form. In modern floras there is known to be a fairly close relationship between mean annual temperature together with mean annual temperature range, and the percentage of plants (woody dicotyledons) in the local flora with entire leaves; an entire leaf is one whose outline is smoothly continuous, as opposed to being lobed or serrated; the margins of the leaflets of compound leaves, likewise, can be described as entire or nonentire. The relationship is illustrated for some modern floras by the graph in Figure 2.7. It should be noticed that the plotted points all refer to sites in eastern Asia, except for two that are in the Hawaiian Islands. The types of vegetation represented range from tropical rain forest, with the highest percentages of entire leaves (up to 86%), through montane forest and warm temperate mixed mesophyte forest, to mixed northern hardwood forest in Manchuria (where only 10% of the species have entire leaves). Arguing back from observed percentages of species with entire leaves in fossil assemblages, Wolfe (1971) concludes that there was a rapid and major climatic deterioration in the Oligocene, and also, probably, pronounced climatic fluctuations in the late Eocene.

A source of error in the "leaf-shape method" of inferring paleoclimatic

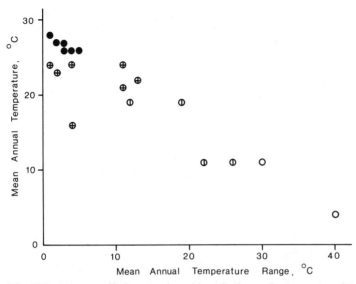

Figure 2.7 The percentage of plant species with entire leaves in the modern floras at 19 geographical locations (all locations are in eastern Asia except two which are in Hawaii): ● > 75%; ⊕ 51% to 75%; ⊕ 26% to 50%; ○ ≤ 25%. (Data from Wolfe, 1971.)

conditions is that the relation between leaf margin and climate shown in Figure 2.7 does not hold on a worldwide basis. Plumstead (1973) has shown that there is a marked difference between the northern and southern hemispheres in the predominant leaf types in temperate forests. In southern latitudes species with entire leaves predominate (typical examples are the eucalypts, oleanders, and *Proteas*) whereas in northern latitudes the majority of woody plants have lobed or serrated leaves (for example, oaks, beeches, elms, hollies, chestnuts, birches, alders, cherries).

Before concluding this discussion on past climates and their effects on plants, a point already mentioned needs to be reemphasized and another introduced. As already mentioned, changing climates cause changes in migration routes; for plants to migrate from one continent to another, more than a land bridge or a row of stepping stone islands is required. Conditions along the route can have strong, but rather unpredictable effects. Species that lack long-distance dispersal mechanisms, and consequently have to "plod" along their migration routes covering only a short step at each generation, must have conditions in which they can thrive and reproduce everywhere along their routes. But wind-dispersed species are affected by climate quite differently. Dense forest forms an impassable barrier to them, but their seeds can easily be transported passively right across savannahs and deserts (hot or cold) to hospitable habitats on the far side. Long-distance wind-powered dispersal of this kind must therefore have been much commoner in dry epochs, when there were many extensive tracts of open country than in wet, when impenetrable rain forests prevailed.

The general conclusion that seems reasonable on the combined effects of evolution and climatic change (since the early Mesozoic) on modern plant distributions (especially of angiosperms) can be summarized in two paragraphs:

1. Primitive angiosperms have inefficient dispersal mechanisms; they are unable to disperse far and cannot escape into new regions if the climate of their ancestral region deteriorates. Therefore they have managed to persist only in regions where the climate has remained much as it was when they first evolved. Southeastern Asia and northeastern Australia meet these requirements (Raven and Axelrod, 1974). Their confinement to unchanging environments also means that they are not exposed to the rigors of strong selective forces; hence they evolve slowly and remain primitive.

2. Advanced angiosperms have efficient long-distance dispersal mechanisms. They have been carried to all parts of the world, and only ice sheets or a total lack of water can prevent their establishment. Since long-distance migrants often find themselves in habitats markedly different from the ones they come from, colonizing populations are subject to intense natural selection. Hence they evolve fast and the gap between advanced taxa and primitive taxa, in evolutionary development, continues to widen.

Some Illustrative Examples

To begin our discussion of the biogeography and paleobiogeography of a few particular plants, it is important to mention *Glossopteris,* a fossil of the Permian period, that is, from a time before the breakup of Pangaea is thought to have started. It was a fernlike plant, probably a true fern, and is so widespread and of such great abundance where it occurs that fossil assemblages dominated by it are described as the "*Glossopteris* flora" (Arber, 1905). This flora has been found in India, Australia, Borneo, South Africa, temperate South America, and Antarctica. Thus its range was Gondwanan and this fact demonstrates that even before the Mesozoic there was a contrast between northern and southern Pangaea, the two halves of Pangaea that were subsequently to become Laurasia and Gondwana.

It is interesting to notice that the contrast still exists. One difference was remarked on above, that plants with entire leaves form a much higher proportion of southern hemisphere than of northern hemisphere floras. Another contrast is that the forests of high latitudes consist entirely of conifers in the north, and of a mixture of conifers and angiosperms in the south. Moreover, in spite of the great antiquity of the gymnosperms, two distinctive families of conifers (that is, of the order Coniferales) are almost restricted to the Gondwana continents; they are the Araucariaceae (the family to which the monkey puzzle tree and the Norfolk Island "pine" belong) and the Podocarpaceae.

Next consider the angiosperms, especially their southern hemisphere* distributions and paleodistributions. Several angiosperm families are represented in all the southern continents except Antarctica (whose climate makes it exceptional). However, there is an interesting contrast between the tropical and temperate floras of the three warm southern continents, Africa, South America, and Australia. In temperate regions the floras of South America and Australia are much more closely related to each other than either is to the flora of temperate South Africa. But in its tropical regions, Africa is not set apart, floristically, from the other two continents.

Good illustrative examples of temperate southern Africa's "odd man out" status are provided by the genus *Nothofagus* and the family Proteaceae. The genus *Nothofagus*, the southern beeches, consists of about 60 species of trees and shrubs, some evergreen, some deciduous. Its present range is South America, New Zealand, Australia, New Caledonia, and New Guinea (see Figure 8.1). But it is absent from Africa. A second example is provided by the family Proteaceae. Its South American and Australasian representatives are much more closely related to each other than either of them are to the South African representatives. Although Australia and South Africa are the two regions where the family is most abundant (there are 35 genera with 800 species in Australasia, and 14 genera with 380 species in southern Africa), no genera are common to these two areas. For instance, *Grevillea* (silky oak), the largest genus, is entirely Australasian, and *Protea* (sugar bush) is entirely African. However, the Proteaceae of Australasia and South America are much more closely related. For instance, the genus *Embothrium* has species in South America, Australia, and New Guinea; and the genus *Lomatia* has species in South America and Australia (Hutchinson, 1964). Although we are here

---

*Northern hemisphere distributions have been more affected by post-Tertiary events than have southern hemisphere distributions. Therefore southern plants are discussed here and northern plants in Chapter 4.

considering angiosperms, it is worth noting that the gymnospermous Podocarpaceae, too, give evidence of the distinctness of temperate southern Africa. The family occurs in all three southern continents; many genera are common to South America and Australia but the family is poorly represented in Africa.

Now consider why Africa should differ from the other two southern continents, especially since it is comparatively close to South America. By far the most likely cause is that Africa separated from the combined landmass of South America, Antarctica, and Australia long before the latter broke up. A land route through Antarctica therefore linked Australia and South America for 50 m.y. or more after Africa became isolated (see Figure 2.2); the climate of this Antarctic land route was temperate. Two other causes may also have contributed to some extent to southern Africa's distinctiveness (Keast, 1973). Africa has drifted farther north than the other two southern continents and it may be that plants it once shared with them have become secondarily extinct in Africa because the change in climate has been greater there. Another possible cause of secondary extinction is that competing plants invading from Eurasia (which has long been in close contact with Africa) may have driven out some taxa that have persisted in South America and Australasia.

Now consider tropical floras. In tropical latitudes, Africa is less of an "odd man out." Among tropical rain forest angiosperms, there are 54 "pantropical" families; and of those that are not pantropical, Africa and Madagascar share more with lands to east and west of them than these lands share with each other (see Figure 2.8) (Axelrod, 1970; Keast, 1973).

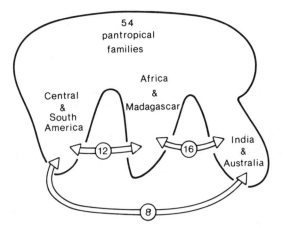

Figure 2.8 Schematic representation of the ranges of families of tropical rain forest plants. (Data from Axelrod, 1970.)

The Cretaceous ancestors of these present-day tropical floras probably belonged to a wide, nearly continuous belt of tropical forest. Raven and Axelrod (1974) believe the angiosperms to have originated in West Gondwana, which was centrally located with respect to dispersal routes leading in all directions. Dispersal was made easy by the fact that the continents, though separating, were still close. Because of the wet, equable climate, the tropical forest zone was wide. The landmasses on which it was centered have drifted north from their Cretaceous latitudes (see Figures 2.1 and 2.5) and are now in the northern hemisphere. It seems likely that the equable climate, the drifting landmasses, and the migrations of the plants themselves gave to many primitive families larger geographic ranges than they have now. Thus many tropical families were able to extend their ranges into the northern and southern temperate zones. Their subsequent history was one of retrenchment, that is, of extinction in regions that became less hospitable. Possible examples of two now southern families that may once have occurred in the northern hemisphere are the Proteaceae and the Epacridaceae (the southern heaths); their fossil pollen has been reported from North America and Britain respectively (Raven and Axelrod, 1972), but the records leave room for doubt.

Indeed, present biogeographic ranges result as much from the contraction of previous ranges as from their expansion. Many families that now occur in two or more of the southern continents probably migrated into them from the north and then became disjunct because of their subsequent extinction in the north.

Plants evolve more slowly, and disperse more rapidly, than mammals. Two consequences follow:

1. Plant families (or other taxa) often have long past histories during which they have had time to travel far; with the passage of time, their geographic ranges have varied considerably.
2. Stemming from this is the fact that, on the average, plant families occur in more than one continent. This contrasts with the mammals, in which well over half the terrestrial families are endemic to a single faunal "region" (roughly equivalent to a continent; see Chapter 1). The proportion of angiosperm families that are endemic to a single continent is much less (Good, 1974) but because of taxonomic disagreements—the splitting of some families and the lumping of others by different taxonomists—it is difficult to say how much less.

One example of an angiosperm family restricted to a single continent is the Xanthorrhoeaceae (the grass trees) of Australia. Endemic to the Americas (North and South together) are the Sarraceniaceae (the insectivorous pitcher plants) and the Bromeliaceae* (which includes pineapples and spanish moss). Indeed, because plant taxa tend to have larger ranges than mammal taxa, phytogeographers tend to subdivide the world into fewer, larger units than do zoogeographers.

3. VERTEBRATE DISTRIBUTION AND CONTINENTAL DRIFT

The past and present distribution of vertebrates has, of course, been determined by continental drift. The fact of drift explains some otherwise puzzling vertebrate distributions, and these we consider first. But drift has controlled the distributions of all organisms, not just a few curiosities, and we consider also what conclusions can be drawn about the past history of a whole class of vertebrates, the mammals.

An important consequence of drift has been that whole faunas have at some times become separated by being rafted apart, and at other times become merged because of the formation between them of an easy dispersal route. The effects of these events on the fate of whole communities are phenomena on which present-day vertebrate distributions can cast some light.

Lystrosaurus, Sphenodon, the Ratites, and the Marsupials

A fossil whose known distribution is of outstanding interest is *Lystrosaurus*, a mammal-like reptile of the early Triassic. Its remains form about 90% of the total in certain fossil assemblages, in which the rest of the fossils are other, less abundant tetrapods (reptiles and amphibians). These assemblages are labeled the "*Lystrosaurus* fauna." *Lystrosaurus* itself was a squat, powerful reptile, about 1 m long, whose only teeth were a pair of downward-pointing tusks; its eyes were high on its head, which suggests it may have spent much of its time submerged with only its eyes above water (Colbert, 1971).

The *Lystrosaurus* fauna has been found at five, now widely separated, sites. Three of these, in what are now India, Antarctica, and South Africa, were, in the Triassic, all in Gondwana (see Figure 2.9). If the animal were known from these sites only, its range would be easy to explain. As Colbert (1971) has noted, a range within Gondwana such as that shown in the figure would not be excessively large for a species of terrestrial reptile; it has about the same area as the modern range of the snapping turtle

---

*One species of the family is found on the West African coast but its presence there is almost certainly the result of chance long-distance dispersal.

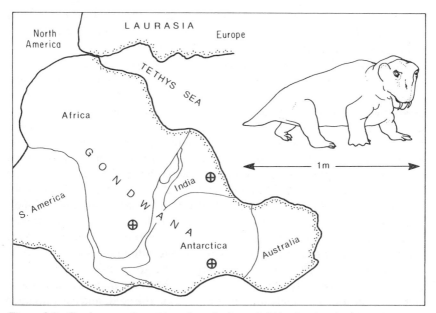

Figure 2.9 Gondwana as it may have been in the early Triassic, showing three sites where fossils of the *Lystrosaurus* fauna have been found. The fauna has also been found in China, which Crawford (1974) believes was a northward extension of the Indian plate and thus part of Gondwana, not Laurasia. (Adapted from Colbert, 1971.)

Chelydra serpentina. However, fossils of *Lystrosaurus* and its associated species have also been found in modern China, in Sinkiang and Shansi, which in the Triassic were assumed to have been in eastern Laurasia. The only possible explanation for this curious distribution, if the lie of the land were as shown in Figure 2.9, would be that some members of the *Lystrosaurus* community had migrated all the way from Gondwana to eastern Laurasia around the head of the Tethys embayment. This seems improbable.

A new, and biologically convincing, explanation of the *Lystrosaurus* distribution has been proposed by Crawford (1974). He considers (see page 34) that Gondwana was larger than it appears in Figure 2.9. According to his reconstruction, the northern edge of the Indian plate, instead of being along the line of the present Himalayas, lay much farther north. Thus a large segment of what is now China (including the sites of the *Lystrosaurus* finds) was part of Gondwana, not Laurasia, and the range of *Lystrosaurus* is no longer surprising.

It is also interesting to note that the occurrence of *Lystrosaurus* in South Africa, India, and Antarctica but not, so far as is known, in

Australia, is one of the reasons for supposing that, before Gondwana disintegrated, India was contiguous with Antarctica rather than with Australia as some reconstructions show it (Hurley, 1968).

Another reptile, primitive but still extant, whose geographic range is puzzling is *Sphenodon punctatus*, the tuatara. It is the only reptile in New Zealand and occurs nowhere else. The problem is not so much why it exists there as why, since it does exist, it is not accompanied by other reptiles and such southern vertebrates as marsupials and monotremes. Raven and Axelrod (1974) suggest the following explanation. In the Cretaceous, New Zealand was much farther south than it is now, in about the 60°S to 70°S latitude belt. Its climate must have been fairly cold (in spite of the fact the world climate was generally warmer in the Cretaceous than now); in addition, winter nights are long in those latitudes. Conditions such as these were probably unendurable for nearly all contemporary mammals and reptiles. However, *Sphenodon* is unusual among reptiles; it has an exceptionally low metabolic rate and remains active at a temperature of 11°C, too low a temperature to permit activity in any other reptile. Probably, therefore, it was the only member of a much richer reptile and mammal fauna able to tolerate the cold, and has remained until the present as sole survivor.

The ratites are another group whose geographic ranges would be extraordinary if the continents had not drifted. These are the large, flightless birds—ostriches, emus, and the like—of the southern continents (see Table 2.2). The presence of different orders on each southern continent could only be explained, if continental drift were disallowed, by one of two theories: (1) The ratites might be unrelated and their resemblance the result of convergent evolution. Or, (2) they might be monophyletic and descended from a flying ancestor; and they might then have dispersed, by flying, among the southern continents, after which each lineage evolved flightlessness independently. Neither of these theories is persuasive; the resemblance among the orders is too close for the first to be credible; the second requires land birds to migrate across wide oceans.

Continental drift makes the current distribution of ratites easily explicable, since one can assume that all migrations were made over land. Cracraft (1973) has suggested the sequence of events shown in Table 2.2. He assumes that all the flightless ratites, and also the Tinamidae, are descended from a common flying ancestor whose home was (probably) western West Gondwana (modern South America); the Tinamidae (tinamous) are weak-flying grouselike birds whose present range is from southern Mexico to southern Argentina. In the Cretaceous, according to Cracraft, two lineages became distinct, the ancestors of the modern tinamous, which retained the ability to fly, and the ancestors of all modern

Table 2.2 Evolutionary sequence in the ratites (family names above; popular names, and modern ranges, in parentheses below. Extinct families are shown by an asterisk. After Cracraft, 1973)

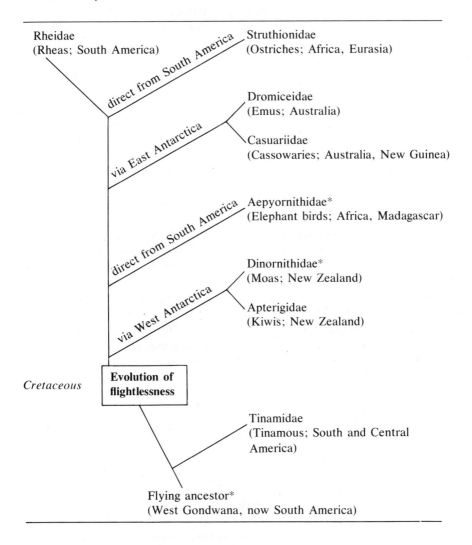

Rheidae
(Rheas; South America)

direct from South America

Struthionidae
(Ostriches; Africa, Eurasia)

Dromiceidae
(Emus; Australia)

Casuariidae
(Cassowaries; Australia, New Guinea)

via East Antarctica

direct from South America

Aepyornithidae*
(Elephant birds; Africa, Madagascar)

Dinornithidae*
(Moas; New Zealand)

Apterigidae
(Kiwis; New Zealand)

via West Antarctica

Cretaceous

Evolution of flightlessness

Tinamidae
(Tinamous; South and Central America)

Flying ancestor*
(West Gondwana, now South America)

ratites, which lost that ability. The table shows the postulated sequence of separations, and the modern ranges of the seven descendent families, of which two have recently become extinct. Also shown are the routes the families are believed to have taken from West Gondwana to their present homes.

The most famous of the exclusively southern groups of vertebrates are the monotremes and marsupials. The monotremes (Monotremata: platypus and spiny anteater) occur only in Australia. They are a very ancient group of unknown history. Darlington (1957) believes they underwent an evolutionary radiation earlier than did the marsupials, and for a time exceeded them in diversity; the few survivors now seem to be approaching extinction.

The marsupials offer more material for controversy. There are three great problems: (1) Where and when did the marsupials originate? (2) How did they reach Australia and what has prevented placentals from following them there and competing with them for possession of the continent? (3) Why are marsupials absent from New Zealand?

At present there are about 170 living species of marsupials in Australia, and about 70 in South America* (Jardine and McKenzie, 1972). The South American species are all small, nocturnal and arboreal.

Inferences as to the place and time of their origin, and their subsequent history, have to be based on fossils, which, of course, reveal nothing about modes of reproduction. The distinguishing features of placental mammals are the morphology of the reproductive system and the retention of the embryo in utero for a period many times as long as the estrous cycle. However, fossil placentals and marsupials are distinguished by their dentition (Tedford, 1974). Modern placentals and marsupials descended from a common ancestor of "marsupioid form" (Cox, 1973). The two lineages diverged no later than the mid-Cretaceous, about 100 m.y. ago. The South American descendents of the common marsupioid ancestor developed into modern marsupials and invaded many other parts of the world from their South American "cradle." The fossil evidence shows that, at latest, they had reached western North America (west of the Midcontinental Sea) by the end of the Cretaceous, Europe by the Oligocene, and Australia by the late Oligocene or early Miocene. Their route from South America to Australia was almost certainly via Antarctica; Australia and Antarctica probably drifted apart in the Eocene (about 50 m.y. ago), and presumably marsupials used the route before it became impassable. The other hypothesis (held by those who do not believe the continents have drifted), that they reached Australia from Asia via the Indonesian archipelago, seems most unlikely (Cox, 1973); if they had come by this route, one would expect at least some of the islands west of

*There are also two in North America, both opossums of the genus *Didelphis* (family Didelphidae). They reached North America from South America after the formation of the land link between the Americas about 6 m.y. ago. Hence they cast no light on the early history of the marsupials.

Wallace's line (see page 14) to contain at least some marsupials to this day, but they do not.

The marsupials eventually became extinct everywhere except Australia and South America (we disregard the North American marsupials which are recent arrivals; see above). They appear to have been unsuccessful in competition with the placentals, which succeeded in excluding them from most of the world. It is surprising, therefore, that the placentals have not invaded Australia and conquered them there too. The probable explanation (Cox, 1973) is that the placentals did not invade South America until after the marsupials had begun their expansion and migration. If the placentals were latecomers, their route to Australia may have been barred by deteriorating climate or even by marine transgressions in Antarctica that sundered the migration routes. A conceivable alternative is that placentals as well as marsupials did invade Australia but, in contrast to the result elsewhere, marsupials became the successful and dominant group in Australia while placentals became locally extinct. In any case, marsupials have thrived in Australia. Australia became completely isolated when it separated from Antarctica and began its drift northward; no further mammal invasions were possible. Moreover, as it drifted northward, the climate became steadily warmer. The result has been a secondary wave of diversification in the Australian marsupials (Tedford, 1974), which now constitute a rich and varied fauna.

In spite of their success in Australia, marsupials are absent from New Zealand. Indeed, except for bats, there are no mammals native to New Zealand, and, until their extinction in the fairly recent past, the flightless moas (family Dinornithidae; see Table 2.2) were the dominant large vertebrates; they were very large, up to 4 m tall. Keast (1972) offers three possible explanations and each has drawbacks. First, New Zealand may have derived its primitive fauna and flora from Australia but have become isolated in the early Mesozoic before mammals originated. Secondly, though both New Zealand and Australia were attached to Antarctica, New Zealand may have drifted away from it too early to receive an influx of marsupials; as Keast remarks (but Raven and Axelrod disagree), the defect of this argument is that a group of ratites (the moas) succeeded in reaching New Zealand and the marsupials are as old as the ratites. Thirdly, it is conceivable that Antarctica was formed of several different landmasses separated by salt water, and New Zealand and Australia were attached to different parts. There appears to be no independent evidence for this possibility; rather, its plausibility stems from the fact that it would provide a good explanation for the extraordinary dissimilarity, in both fauna and flora, of Australia and New Zealand.

The Paleobiogeography of Placental Mammals

The most advanced vertebrates to have evolved so far are the placental mammals. The earliest mammals (before placentals and marsupials became differentiated) probably appeared at about the time of the Triassic-Jurassic transition (Kurtén, 1972); the two lineages diverged later, perhaps in the mid-Cretaceous, and their greatest diversification was in the late Cretaceous and early Tertiary. The placentals replaced the giant reptiles (the dinosaurs) fairly rapidly as the dominant land vertebrates and have maintained that dominance ever since, through all the Tertiary and up to the stage we have now reached in the Quaternary (equivalently, through all the Cenozoic so far).

The place of their origin is uncertain. Darlington (1957) considered the Old World tropics to be their evolutionary center. Cracraft (1973) pointed out that because warm climates and tropical rain forests extended into much higher latitudes at the time the mammals were diversifying than they do now, there need not have been a limited evolutionary "center"; instead, evolution could have taken place over a large area. For most of the primitive placental orders, this area was probably Laurasia.

Kurtén (1969) attributes* the great diversity of early Tertiary mammals to the fact that the earth's land surface was at that time broken up into eight more or less isolated segments. The breakup of Pangaea had culminated in the formation of seven continents (North America, Eurasia, India, South America, Africa, Australia, Antarctica) and for a period before plate collisions created new joins, these were separated. Further, Eurasia was split into two parts by the Turgai Strait, which stretched from the Tethys Sea to the Arctic Ocean. Thus as many as eight semi-isolated "nuclear areas" may have existed simultaneously. This permitted the evolution of no fewer than 29 orders of placental mammals, of which 13 have since become extinct. Thus disregarding the 2 marine orders (Cetacea: whales, and Pinnipedia: seals), there are now only 14 orders of terrestrial placentals and the landmasses they occupy have been reduced from eight to four. This reduction has come about because of the glaciation of Antarctica, the joining of the Americas, and the joining into what is effectively one great northern continent of North America, Eurasia (including North Africa), and India. The four present landmasses are therefore this "northland" which Kurtén calls the Holarctic-Indian region, the Neotropical region (South and Central America), the Ethiopian region (Africa south of the Sahara), and Australia.

*For a dissenting view, see Charig (1973).

Only the first three of these contain placental mammals. The places of origin of the 14 extant orders are believed to be as follows (Kurtén, 1969).

Holarctic (Laurasia)	Africa
Insectivora (shrews, moles, etc.)	Proboscidea (elephants)
Primates (lemurs, monkeys, apes, etc.)	Hyracoidea (hyraxes)
	Tubulidentata (aardvarks)
Rodentia (rats, mice, squirrels, etc.)	Sirenia (manatees and dugongs)
Carnivora (wolves, weasels, lions, etc.)	
Perissodactyla (zebras, tapirs, etc.)	South America
Artiodactyla (pigs, deer, antelopes, etc.)	Edentata (sloths, armadillos, etc.)
Chiroptera (bats)	
Lagomorpha (rabbits, hares)	
Pholidota (pangolins)	

Thus a halving (from 27 to 14) of the number of orders of terrestrial placental mammals appears to have accompanied the halving (from eight to four) of the number of the world's distinct landmasses. We now consider whether this phenomenon is a case of cause and effect.

Merging Faunas and Competitive Exclusion

Kurtén (1969) considers that the steady attrition in the number of extant mammal orders since the early Tertiary is a direct result of the amalgamation of previously separate landmasses. The process of competitive exclusion, familiar to ecologists as a small-scale, short-term phenomenon, is assumed to have taken place on a worldwide scale, over geological time spans.

Many studies have been made of the details of competitive exclusion in particular regions. A well-known example is the extinction of many families of mammals in the Pliocene following formation of the Isthmus of Panama (Simpson, 1950, 1969; Webb, 1976). Before a land link had formed, North and South America had no families of terrestrial mammals in common. Once a migration route came into existence, a faunal interchange took place and the fauna of each continent was temporarily enriched by immigrants from the other. About 15 families spread from North

America into South, and 7 in the opposite direction. The increased diversity did not last, however. Many families became extinct and now there are about as many in each of the two continents as there were before the interchange began, about 25 families in North America and about 30 in South America. But the composition of the South American fauna has been completely altered. Many families that were endemic to South America in the early Pliocene are now extinct and their places have been taken by invaders from North America such as various ungulates, rodents, carnivores, and rabbits. The faunal richness of the Americas as a whole has been drastically reduced.

It has been argued that to ascribe these extinctions to competitive exclusion involves a circular argument. Those who hold this view apparently think that "competitive exclusionists" first infer, from observed extinctions, that exclusion is going on and then use these same extinctions as evidence of the correctness of their hypothesis. Though this may be a logically objectionable procedure, it has its merits when one considers the alternative, which is to deny the reality of a cause-and-effect relationship between two events that, experience shows, repeatedly happen in the same order; these events are the joining of landmasses and the extinction of animals. In debates between competitive exclusionists and their opponents, the burden of proof surely falls on the latter. After all, in paleobiogeography, all conclusions about processes must be based on uncontrolled natural experiments that took place several million years ago.

In comparing biogeographic and ecological cases of competitive exclusion, an important contrast to note is that in ecological contexts the victim is usually a species, but in biogeographic contexts it is a larger taxon, a family or even an order that becomes extinct, locally or globally. For example, in South America *all* species of the order Litopterna (pseudohorses with short trunks, known only from South America) vanished and their place was taken by such North American Perissodactyla as horses and tapirs. There was not an exchange that left some Litopterna species as survivors in each continent. It follows that families and orders *as a whole* are strong or weak in competition.

It is therefore worth considering the following problem. Suppose two species (say A_1 and A_2) from family A, and two species (B_1 and B_2) from family B all find themselves in one continent (whether in one habitat type we cannot tell); and suppose B_1 and B_2 become extinct, while A_1 and A_2 survive. Two contrasting explanations might be entertained. (1) Perhaps family A as a unit excluded family B as a unit. (2) Perhaps A_1 excluded B_1 and A_2 excluded B_2 and the reason the victor in both cases was a member of family A arose from the superiority of the family, a superiority enjoyed by each of its member species individually. Thus two distinct modes of competitive exclusion at the superspecific level are at least theoretically possible. Comparative investigations are needed.

It is also worth noticing that if species A_1 invaded the territory of species B_1 (when intermigration became possible) but not vice versa, one possible inference is that population growth in species A_1 had previously been kept in check by a density-dependent mechanism; removal of this check permitted A_1 to increase numerically and hence to expand its range into the newly accessible area. Species B_1, on the other hand, might previously have been controlled by a density-*in*dependent mechanism and therefore did not enlarge its population and hence its range when new territory became available.

Another, quite different, explanation of "lopsided" migration (A_1 into B_1's range but not B_1 into A_1's) might be that A_1's ecological amplitude exceeds B_1's.

It is intellectually unsatisfying to ascribe extinctions to competitive exclusion and let it go at that. The way in which exclusion operates, and between what competing entities, deserves careful study. It is interesting to consider whether cases can be found where we can observe competitive exclusion actually in progress. Two possible examples have been described by Darlington (1957), both among birds.

The first concerns long-legged birds, both waders and ground birds. Among long-legged waders, the dominant family is the Ardeidae (herons, bitterns, egrets). At least six genera and two species of the family are cosmopolitan. Darlington believes the Ardeidae may be replacing such other long-legged birds as the Gruiformes (cranes, rails, bustards, limpkins) and the Ciconiidae (storks and ibises). Some members of these two groups live away from water but this may be "a result rather than a disproof of competition" (Darlington, 1957).

As a second example, Darlington points to the seeming decline, in the modern world, of two old and complex, related families, the Columbidae (pigeons and doves) and Psittacidae (parrots). These families are dominant in many tropical communities. However, parrots no longer occur in the north temperate zone although they did in the Miocene, in France and Nebraska. The interpretation of other evidence implying the decline of these families depends on whether or not one believes that the continents drift (Darlington did not). In any case, Darlington is of the opinion that the passerines (order Passeriformes: the 'songbirds', the majority of north temperate land birds) may be in the midst of replacing pigeons and parrots in many parts of the world.

An aspect of (apparent) competitive exclusion that requires careful investigation in every case is the following (for conciseness, it helps to describe it in teleological terms). Can one assume that when competitive exclusion appears to be happening, one species is always an "aggressor" excluding a "victim" that in the aggressor's absence would have persisted? Could it not in some cases be true that the "loser" was doomed in any case, and the "winner" is just availing itself of resources left unused by the dwindling population of losers? Of course, this cannot explain the extinctions that so frequently follow the merging of hitherto separated faunas.

4. INVERTEBRATES AND CONTINENTAL DRIFT

Comparatively little work has been done relating continental drift and the distributions of terrestrial invertebrates, because the great majority of terrestrial invertebrates leave no fossils. However, some interesting attempts have been made to infer the order of separation of the continents from knowledge of the present distribution of related groups of organisms, and, arguing the other way round, to infer the antiquity of modern species from assumptions as to the times of continental separations. This section mentions a few representative examples of these researches.

Schminke (1974) bases his arguments on the distributions of some very small (1 or 2 mm long) malacostracan Crustacea of the family Parabathynellidae that live buried in the sand at the bottom of fresh water. The most primitive members of the family are found in eastern Asia, which is treated, on that account, as their "center of evolution." Two separate lineages are recognizable and they radiate (in a geographic sense) from the point of origin; one "line" leads west, to Europe, Africa, and South America, the other south to Australia, reappearing again in South America. Along each line a series of genera are encountered in which the morphological characters become steadily more advanced as distance from the origin increases. The observations can only be accounted for satisfactorily, according to Schminke, if we assume the existence of Mesozoic land connections joing Asia to Africa, Madagascar, and South America in a westerly direction, and to Australia, New Guinea, and New Zealand with a route to South America via Antarctica in a southeasterly direction.

Freshwater organisms like these crustaceans, that cannot endure exposure to salt water or drying, have very restricted powers of dispersal. It is difficult to imagine their being successfully transported across the sea; their delicacy makes it unlikely they could be transported by birds, and they are not adapted to dispersal by wind. Therefore, the occurrence of a single species, or genus, on either side of an ocean prompts the inference that the disjunction was caused by continental drift. Two other such examples will be mentioned, one dealing with freshwater planaria (phylum Platyhelminthes, flatworms; class Turbellaria) and the other with soil nematodes (phylum Nematoda, roundworms and threadworms).

Planarian biogeography has been analyzed by Ball (1975). He first considers subgenera of the genus *Dugesia* and, on the basis of their modern distributions, hypothesizes that they dispersed northward from a point of origin in that part of Gondwana that is now Antarctica, and that they did not reach the northern hemisphere until after the Atlantic had formed, making dispersal between the Old and New Worlds impossible. Having arrived at a reasonable hypothesis on the basis of the ranges of the subgenera of *Dugesia*, Ball then sought to test the hypothesis by examining the ranges of two other, more advanced, planarian genera, *Polycelis* and *Planaria*. He sought to answer the question: do the ranges of *Polycelis* and *Planaria* support or falsify the hypothesis based on *Dugesia*? At first glance they appear to falsify it, since both genera have amphi-Atlantic ranges (see Figure 2.10). However, Ball considers a more likely explanation of the seemingly discrepant observations to be that both *Polycelis* and *Planaria* are diphyletic (or even polyphyletic), and that separate lineages of each genus followed the same diverging northward routes from ancient Antarctica, on either side of the proto-Atlantic, as did the subgenera of *Dugesia*.

McDowall (1973), discussing earlier work by Ball, argued that transoceanic dispersal of these planarians cannot be ruled out since *Dugesia* also occurs in the very isolated Crozet Islands in the far south Indian Ocean (see Figure 8.6). If, as McDowall asserts, the islands are young, *Dugesia* could only have reached them by long-distance dispersal over salt water and its presence there would prove that, difficult or not, such dispersal must be possible. But Ball (1975) considers that the Crozet Islands are isolated fragments of Gondwana and that this accounts for *Dugesia*'s presence in them. It is unclear how the genus survived if the islands were glaciated in the Pleistocene.

The biogeography of soil nematodes has been studied by Ferris, Goseco, and Ferris (1976). The genera concerned, *Dorylamoides* and *Tyleptus*, are free-living nematodes of freshwater and soil. Their powers of dispersal are poor, as they must always be in contact with water or moist soil. Ferris and co-workers wished to learn about the nematodes, in particular their evolutionary rates, from a knowledge (taken as given) of continental drift and of the species' modern ranges. Their approach contrasts with that of most earlier workers who have used biological data as

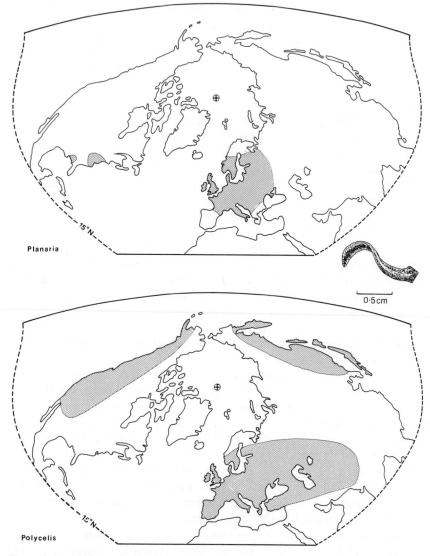

Figure 2.10 Modern distribution of the planarian genera *Planaria* and *Polycelis* (Planariidae). (Redrawn from Ball, 1975.)

evidence of drift rather than letting the known facts of drift yield new biological knowledge. Their conclusions are that a primitive* species of *Dorylamoides*, *D. leptura*, that is now found in India and South America, must have evolved before Gondwana split up and must, therefore, have existed as a species for at least 100 m.y. This is an extraordinarily long time for a species to persist. It is instructive to compare it with, for example, the equivalent time span in insects, about which Ross (1972) has written, ". . . it takes at least 30,000 years of isolation for [an insect] population to evolve into a species that is genetically distinct from other segments of the parental species." The period for nematodes appears to be on the order of 10,000 times as long.

All species of the other nematode genus studied by Ferris and co-workers, *Tyleptus*, are evolutionarily fairly advanced. Since they are present in India, Africa, and the Americas, but not in Eurasia or Australia, it is inferred that the evolutionary radiation of the genus cannot have occurred earlier than about 100 m.y. ago. Thus the relative ages deduced for these nematodes, on the basis of their modern geographic ranges, accord with their relative levels of morphological advancement.

Although long-distance dispersal is much easier for aerial than for freshwater organisms, the distributions of many small insects and spiders is more likely the result of continental drift than of transoceanic journeys. Platnick (1976) has suggested that two related spider genera of the subfamily Laroniinae with nonoverlapping ranges (one genus occurs only in the Laurasian continents, the other only in the Gondwanan continents) became differentiated after the range of their common ancestor was split in two when Pangaea broke up.

Detailed studies have been carried out by Brundin (1967) on the biogeography of the chironomid midges [order Diptera (Nematocera), family Chironomidae] of the southern hemisphere. He bases his arguments on the supposition that pairs of species with ranges now widely separated are recognizable as "sister" species, that is, as immediate descendents of a common ancestral species (see page 68). On the basis of the pairs thus recognized, Brundin presented a detailed history of Gondwana. According to his reconstruction, the course of events was as follows. In the late Jurassic Gondwana had a long coastal mountain range, which has now become New Zealand, western Antarctica, and western Patagonia. The mountains offered a great variety of habitats and climates and fostered great evolutionary diversification in the Chironomidae.

---

*The primitive character of the species is that the female is didelphic, that is, has two equal gonads. Advanced species are monodelphic, with only the anterior gonad of the female developed.

Chapter Three

EVOLUTION, PHYLOGENY, AND GEOGRAPHIC SPREAD

Evolutionary changes and geographic range changes befall a species simultaneously; each process affects, and is affected by, the other. Indeed, the two kinds of change are inseparable and should always be thought of together.

The ultimate aim of evolutionary research is to reconstruct the phylogeny of all known species, extinct and extant; that is, to reconstruct the phylogenetic "family tree" which shows the pattern of descent of all species from an ultimate common ancestor. The way this may successfully be done is at present (the 1970's) a hotly contested topic. The two chief approaches are fundamentally different in underlying philosophy even if the results they yield often resemble each other. They form the subject of this chapter, particularly of the first three sections.

Section 4 discusses a well-known biogeographic phenomenon, the pronounced latitudinal gradient in the species-richness of floras and faunas, commonly called (for brevity) the "latitudinal diversity gradient."

Section 5 is a mathematical section in which are derived the logical consequences of some simple assumptions about the geographic ranges of related species.

1. PHYLOGENETIC AND EVOLUTIONARY SYSTEMATICS

Systematists—those who attempt to classify all organisms in such a way that their genetic interrelationships are displayed—are divided into two schools or even, one could say, into two warring groups. Before considering the biogeographic implications of their debate, it is important to make as clear as possible the underlying causes of the fierce disagreement between them. A fair and clear comparison of their views has been given by Cracraft (1974) and his account has been used as a basis for the summary below. His paper should be consulted for details and his bibliography for references.

The contrasted systems will here be given the names which their proponents favor, "phylogenetic systematics" and "evolutionary systematics," though, as we shall see, they have other names as well.* The numbered pairs of paragraphs below summarize the properties (paragraphs 1 and 6) and the main tenets (paragraphs 2, 3, 4, and 5) of the two schools. Paragraphs 1A, . . . , 6A refer to the phylogenetic school; paragraphs 1B, . . . , 6B, refer to the evolutionary school. Below each A/B pair, and inset to distinguish it, is a paragraph or two of additional explanatory comment. To conclude the section a number of quotations from the literature are given which show how deeply convinced some members of each side are of the unassailability of their position. Before considering the numbered paragraphs, it should be remarked that the assigning of paragraphs labeled A to phylogenetic systematics and of paragraphs labeled B to evolutionary systematics should not be taken to imply that the former system is superior to the latter. The order was chosen for clarity of exposition.

1A. *Phylogenetic systematics is also known as "phylogenetic cladistics," "cladistic systematics," or simply "cladistics." Among its chief proponents are Hennig, Brundin, Nelson, and Cracraft.*

1B. *Evolutionary systematics is also known as "morphological systematics" or "phenetic systematics." Among its chief proponents are Simpson, Mayr, Bock, and Ghiselin.*

> The name "cladistics" derives from the Greek "klados," a branch or twig. "Phylogeny" is from the Greek "phylon," a race or tribe. "Phenetics" is from the Greek verb "phainen," to appear. In what follows the two kinds of systematists are referred to as cladists and pheneticists for short.

2A. *The ancestors of known organisms cannot be determined with certainty. Postulated, hypothetical ancestors are not to be equated with known fossils.*

---

*Each side arrogates to itself a name indicative of generality and gives the opposing side a name stressing its alleged bias. These names are cogently discussed by Bock (1969).

2B. *Direct, empirical knowledge about ancestors is accessible to paleontologists.*

Cladists consider that to infer direct ancestor-descendant relationships between known fossils and extant organisms is to make untestable assumptions that, more often than not, are misleading, in fact lead to the construction of faulty phylogenies. Pheneticists normally search for the ancestors of extant species among known fossils. The relationships they infer are usually that an extant family or genus is descended from a known fossil family or genus; some pheneticists believe they can recognize lineages of species-level taxa.

3A. *The splitting of a species-population into two or more subpopulations that then diverge and evolve into new species is the ONLY mode of origin of new species.*

3B. *There is another process that leads to the formation of new species. Because of slow, continuous evolutionary change, organisms often differ so greatly from their remote ancestors that one cannot treat earlier and later members of a long lineage as members of the same species.*

Cladists do not deny that the slow transformation of one taxon into another does occur, but they regard it as of negligible importance compared with speciation by splitting. The slow transformation of taxa is often called "phyletic gradualism" ("phyletic," like "phylogenetic," is from "phylon," a race; see above). A new species formed thus gradually is sometimes called a "chronospecies." There is, of course, no precise instant at which a species undergoes conversion to its descendant chronospecies and, of necessity, the only perceivable differences between succeeding species are morphological. Since, by definition, their member individuals never live at the same time, it is meaningless to ask whether or not they are interfertile; the opportunity for hybridization can never arise. It should be noticed that, for both schools of systematists, tenets 2 and 3 are linked. Cladists deny that ancestors of living species can be reliably recognized in fossil forms, and hence do not construct long lineages that could exhibit gradual change. Pheneticists assert that sequences of fossils that look as though they form true lineages are probably true lineages in fact, and the changes along them are (in part) manifestations of phyletic gradualism (the words "in part" are to emphasize that splitting of lineages is assumed to be going on as well).

4A. *The branching in a phylogenetic tree must always be dichotomous.*

4B. *When a species-population splits into reproductively isolated sub-populations that then diverge to become new species, a multiple (polytomous) split is not improbable, hence polytomous as well as dichotomous branchings can appear in a true phylogenetic tree.*

Notice that for tenet 4, as well as for 2 and 3, the differences of opinion between cladists and pheneticists are not absolute. In each case cladists assert only that an hypothesized event (discovery of a true ancestor, evolution of a chronospecies, multiple splitting of a parent species), though theoretically possible, is so improbable that if it is allowed to govern construction of a supposedly true phylogenetic tree, that tree is almost certainly erroneous. Notice, also, that cladists do not deny that a species-population may often become separated into more than two isolated subpopulations simultaneously. What they rule out as excessively improbable is the attainment of specific rank by these several isolates all at the same time.

5A. *The only way to construct a phylogenetic tree is to join pairs of "sister" groups. These are the two groups that are sole descendants of an hypothesized immediate common ancestor whose dichotomous splitting produced them.*

5B. *Phylogenetic trees are to be constructed by joining groups whose close relationship is inferred from their overall similarity.*

The joining of recognized "sister" pairs is at the heart of the cladists' method, and the results are displayed as "cladograms" which are of the forms shown in Figure 3.1 or Figure 3.2a. The salient common property of these representations is that there are no labeled entities anywhere except at the top where branches end. This is because the junction points where branches join represent no more than hypothetical, unrecognized ancestors, as opposed to known fossils (see paragraph 2A). Each joining of a pair of sister groups creates a "monophyletic group," defined by Hennig (1975) as a group of species that have an ancestor—a "stem species"—common only to themselves. Thus the following is an exhaustive list of the monophyletic groups in the cladogram in Figure 3.1, left: (1, 2), (3, 4), (5, 6), (7, 8), (1, 2, 3, 4), (5, 6, 7, 8), and (1, 2, 3, 4, 5, 6, 7, 8). All other subsets are nonmonophyletic.* Sister pairs are

---

*Nonmonophyletic groups may be paraphyletic or polyphyletic, or a mixture of the two. The meanings of these terms are not discussed here as they will not be needed in what follows. Ashlock (1974) defined them in a review paper, but his definitions were subsequently

recognized from the advanced characters they have in common, not the primitive characters. This very important point is best illustrated by a deliberately absurd example, chosen to illustrate the distinction. Tigers and gorillas are both hairy and hence, presumably, related; but they differ from each other in the anatomy of their forefeet. Gorillas and human beings both have opposable thumbs and hence, presumably, are related; but they differ greatly in hairiness. Should they be classified as on the right below, or as on the left?

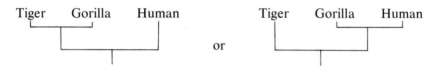

The obvious absurdity of the tree on the left arises from the fact that it attributes a closer relationship to the pair sharing the primitive character hairiness (which is common to all mammals except for a few that are secondarily almost hairless) than to the pair sharing the advanced character, possession of an opposable thumb. It is necessary here to introduce another two technical terms that will be needed in what follows. Primitive characters, such as hairiness in mammals, are known as *plesiomorphous* (from the Greek "pleio," majority, and "morphe," form or shape, combined to mean a widespread, unspecialized, form). Advanced characters, such as the opposable thumb in Primates, are *apomorphous* (from the Greek "apo-," from, or out of, in the sense of "derived").

Pheneticists construct phylogenetic trees by uniting related groups which are recognized on the basis of their overall similarity. There is no objection to forming groups with more than two members. The trees are of the form shown in Figure 3.2b with labeled taxa placed at ancestral positions in the tree if it seems appropriate to put them there. Thus fossils may be regarded as ancestral chronospecies (species 1 and 2 in the lefthand tree and species 1 in the righthand tree), as species that split prior to speciation (species 2 in the middle and species 2 on the right), or as species that have left no descendants (species 1 in the middle tree). Hypothetical ancestors to which no known fossils correspond are interpolated where needed. Construction of a tree depends on both "vertical" and "horizontal" relationships and all available evidence is taken into account in judging such relationships; that is,

contradicted and corrected by Hennig (1975), the originator of the terms. It may be useful to know that Ashlock suggests the word *holophyletic* as a synonym for Hennig's *monophyletic*.

attention is not restricted to so-called apomorphous (or advanced) attributes.

6A. *According to its proponents, phylogenetic systematics is a simple, logical way of constructing phylogenies. The method has an axiomatic basis and clearly defined rules. It is neither subjective nor imprecise.*

6B. *According to its proponents, evolutionary systematics avails itself of the experience and judgment of its practitioners who take account of all relevant evidence in constructing phylogenies. They do not deliberately and wastefully ignore pertinent facts because to consider them would break some self-imposed rule.*

This final A/B pair of paragraphs shows that the debate between the two schools really hinges on whether, in constructing phylogenies, one should admit or exclude so-called "intuition," that is, extraneous, unstructured bits of knowledge one happens to possess, which combine to constitute experience. Cladists disregard experience on grounds of logical rigor. Pheneticists admit it on the grounds that not to do so is to waste relevant information.*

The foregoing paragraphs summarize the debate. As may be seen, the disagreement is over methods of achieving a desired result, the construction of true phylogenetic trees. Ultimately, there would be only a single tree, incorporating all living things without exception (assuming life originated only once); in practice, work concentrates on more modest, less inclusive trees, intended to show the relationships within the taxon—a family or subfamily, say—that a particular systematist is working on. Also, the systematist's objective is usually to devise a classification simultaneously with a phylogenetic tree; the classification is required to be useful, as well as to be "natural" in the sense of displaying the true relationships of the species classified.

Although these motives activate systematists of both schools, their priorities differ. Cladists wish to construct phylogenies and classifications in accordance with certain rules, designed to prevent the illogical jumps to wrong conclusions of which they accuse pheneticists. Pheneticists wish to

---

*This dilemma is not restricted to biological systematics. It is encountered in all sorts of contexts. For example, should a driving test examiner pass or fail a candidate who performed excellently in the test, but whom the examiner recognizes as the same person who drove through a stop sign and almost collided with him that very morning. A cladist would say "pass," a pheneticist, "fail."

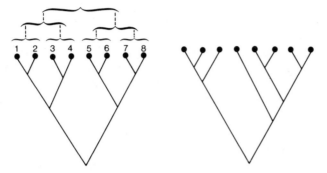

Figure 3.1 Two cladograms. Observe that all branching is dichotomous and that recognized taxonomic groups occur at the final branch tips only. The brackets over the cladogram on the left show how the taxa combine to form monophyletic (holophyletic) groups. Further details in text.

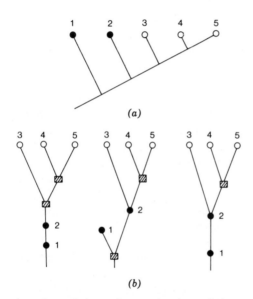

Figure 3.2 Two ways of postulating and portraying the evolutionary relationships among five taxa. Black dots are extinct taxa known only from fossils; hollow dots are extant taxa; shaded squares are presumed, undiscovered taxa. (a) A cladogram of the simplest possible form; black dots and hollow dots are all in the top tier and there are no shaded squares. (b) Three of the many conceivable phylogenetic trees that conform with the cladogram. (Adapted from Szalay, 1977.) Further details in text.

understand the course of evolution (and embody the result in a classification) by whatever means lie to hand and are unwilling to make the task needlessly difficult by imposing what they regard as arbitrary rules.

It is difficult to describe the debate impartially. Most commentators do not even try. The opinion of a person (myself) who is a consumer rather than a creator of phylogenies and whose opinion, therefore, is probably unsought, is that the pheneticists have a slight but significant lead. My reason for this opinion is best explained by the following analogy. Anybody who frequently uses the identification keys in published "floras" and "faunas" in order to identify unfamiliar plants and animals knows that rigorous (or, to use an emotive word, slavish) adherence to the path that ought to lead to the answer occasionally yields an absurd result. Such an outcome can only be avoided, when it arises, by allowing experience (extraneous knowledge plus common sense) to reveal the fault in a path whose logical perfection was an illusion. To repeat, the foregoing is *only* an analogy.

Nothing can give a better idea of the intensity of the debate than a few quotations.

Thus according to Brundin (1967), in constructing phylogenies, "adherence to strict monophyly must be inflexible. This demand means . . . that the monophyletic status of even plesiomorph (primitive) groups must be demonstrated by the application of apomorph characters." And later: "All attempts . . . to interpret the history of life without . . . consideration of these things are thus futile." From Brundin (1972a): "The evolutionary process is . . . orderly . . . . [Its] course is determined by the interaction of two grand directional forces [and] there are but narrow margins left for chance." These forces are "pre-determined species- and group-specific potentials, and . . . natural selection." Hennig (1975), in a paper devoted wholly to arguing the debate, writes inter alia of the "logical demand for strict uniformity of viewpoint. . . . [R]ejecting a logical demand . . . is . . . a reckless practice." And later in the same paper: "The logical priority of the phylogenetic system . . . arises from its foundation in a biological theory with unambiguously defined central concepts."

Turning now to the pheneticists' side, it is worth quoting Szalay (1977) who writes of "cladistic dogma" and "an unrealistically rigid superstructure" and "this subjugation of a practice of systematics . . . to theoretical concepts which are artificial and inapplicable." Further "the use of cladograms alone obscures those hypotheses of phylogenetic relationships that are proposable and between which reasoned choices can be made." Notice that Figure 3.2 is adapted from a figure in Szalay (1977). He argues that often it is not only possible, but biologically very instructive, to make a reasoned choice among (for example) the three trees in the bottom of the figure. These trees embody testable assumptions, and knowledge is acquired by testing assumptions. To insist, as cladists do, that interrelationships among species must only be portrayed in cladograms amounts to a refusal even to consider interesting, and perhaps testable, hypotheses concerning ancestor-descendant relations.

Sokal (1975) also is critical of cladograms. He writes, "By restricting their classifications to cladograms, cladists are removing from their classifications a potentially interesting and crucial aspect of macro-evolution—the invasion of new adaptive zones." (This would lead to evolution of chronospecies.)

Classifications based on phylogenetic (cladistic) systematics often yield results that look wildly aberrant to traditional evolutionary systematists. The hierarchical level that a cladist assigns to a monophyletic group—a taxon—depends solely on how many tiers of junction points in a clado-gram separate the taxon to be named from the topmost tier of empirically observed species. As Byers (1969) puts it, "Identification of . . . sister groups for the purpose of constructing a phylogenetic classification is made difficult by the fact that in other existing classifications they may have been assigned to different categorical [= hierarchical] levels." For instance, Hennig treats the Mammalia and the Diptera as of equal rank; in evolutionary systematics, the Mammalia are a class, and the Diptera only an order within a class, the Insecta; continuing, Hennig treats the placental mammals (in evolutionary systematics, a subclass) as of equal rank with an insect family.

Members of the two schools are unlikely ever to be reconciled. An interesting psychological study could be made contrasting the two sets of antagonists. There is more to say about these clashing opinions in Section 2, where we concentrate on specifically biogeographic facets of the debate.

2. EVOLUTION AND DISPERSAL

When we consider the biogeographic aspects of evolutionary theory, it is found that the number of disputing schools has gone from two to three. They are as follows.

School 1 makes three assertions: (i) that all groups tend to speciate most actively in a limited area that constitutes a center of origin; (ii) that when an ancestral species splits to yield two daughter species, one is always evolutionarily more advanced than the other; the species pair (which is a "sister" pair) consists of one apomorphous (advanced) member and one plesiomorphous (primitive) member; and (iii) that the plesiomorph descendant continues to occupy the original geographic range of the ancestor while the apomorph descendant moves to a new, peripheral region, away from the center of origin. The chief proponents of this school are Hennig and Brundin.

School 2 agrees with school 1 with respect to tenets i and ii but holds an exactly contradictory opinion in place of iii. According to school 2, after a species has split it is the apomorph descendant that occupies the ancestral region (the center of origin), and the plesiomorph descendant that moves to the periphery. The chief proponent of the school is Darlington.

School 3 asserts that the concept of a center of origin is inconsistent with the concept of allopatric speciation.* Hence apparent centers of origin are illusory and biogeographic theory should rid itself of the notion. The proponents are Croizat, Nelson, and Rosen.

There is scope here for three two-way disputes and a three-way free-for-all. However, a survey of the battleground shows two unconnected engagements in progress: the first is between school 1 and school 2; the second an attack by school 3 on schools 1 and 2 combined. Let us consider the debates in turn, in the order given.

The Center-of-Origin Concept

While there is undeniably much force in the argument of school 3, namely that the center-of-origin concept is difficult to reconcile with that of allopatric speciation, it is equally undeniable that centers of origin appear to be common. When isopleth ("contour") maps are constructed in which the isopleths join points of equal species-richness or genus-richness (within some taxonomic group such as a family), it is frequently found that the map shows one or two "hills" covering regions in which the number of species or genera reaches a peak. This phenomenon is very common. Figure 3.3 shows examples and numerous others may be found in, for example, Stehli (1968) and Good (1974).

We defer to Sections 3 and 4 a discussion of how, assuming speciation is never sympatric (see page 82), species densities can become high within "centers" of restricted area. For the present, assume that this does happen, and also that all species splits result in the evolution of one apomorphous and one plesiomorphous daughter species. Now consider

---

*The evolutionary divergence to form new species of geographically separated fragments of a once-continuous parent population (from the Greek "allos," different, and the Latin "patria," native land). Allopatric speciation is believed by many students of evolution to be almost the only method of formation of new species, with very few exceptions; but see page 83.

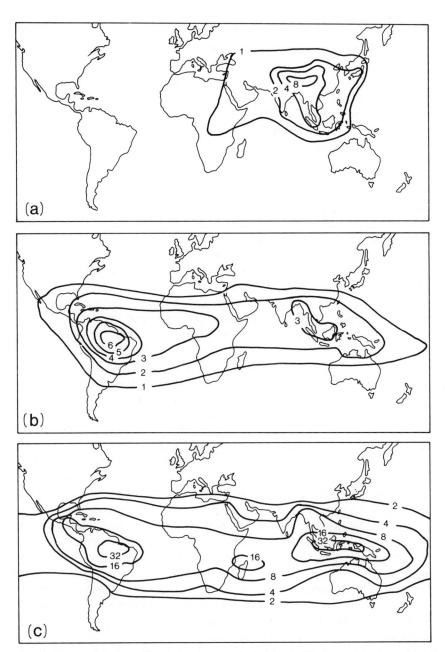

Figure 3.3 Centers of origin of (*a*) genera of pheasants (Phasianinae); (*b*) species of Crocodilia; (*c*) genera of palms (Palmae). The isopleths join points of equal taxon density. (Adapted from Stehli, 1968, whose maps show the control points, and where references to the data sources are given.)

the disagreement between schools 1 and 2. Here are some quotations from the leaders.

For school 1, Brundin (1972) writes, "Critical biogeographic research has to accept as a guiding principle that the primitive group at least primarily is closer to the area once occupied by the ancestral species than the derivative sister group." Also, "Adherence to conservatism, as signified by comparative plesiomorphy in one of two sister species taking over the larger part of the ancestral gene pool, means insurance of persisting possibilities for breakthrough of the evolutionary potential of the ancestral species if the experiment symbolized by the derivative, apomorphic sister species proves unsuccessful." This extraordinary teleological argument seems to imply that the plesiomorph species consciously adopts a wise course of action.

In an earlier paper, Brundin (1967) discusses how, assuming the Hennig-Brundin set of tenets to be correct, one may infer the direction of spread of related groups of species. One way is to apply the "rule of deviation" according to which if the two daughter species of a single ancestor diverge, both evolutionarily and spatially, "the marginal one becomes more or less conspicuously apomorph" while the other, which remains in the area of origin, "remains more or less plesiomorph." "This is the rule of deviation. [It is valid for] single characters and the general design alike, for sister species as well as for groups of any rank." To dispel any remaining doubts on this school's opinions, here is a final quotation (the italics are mine): "The *only* method which is able to promise *conclusive* evidence in biogeography . . . [was] set out by Hennig sixteen years ago. . . . The task of the potential biogeographer *obviously* is to try to reconstruct the actual sister group systems. The phylogenetic relationships *have to* be worked out by means of apomorph (derivative) characters they have in common, the *rule* of deviation, and the *rule* of geographical vicariation* of sister groups." It will be seen that biogeographic reconstructions by the Hennig-Brundin method require one to recognize which is the plesiomorph and which the apomorph, given a pair of sister species. In some cases this is easy, in others, as Byers (1969) expressed it, "It is not altogether clear how this is accomplished." Or as Edmunds (1972), who worked with Ephemeroptera (mayflies) warned, "Determination of which character state is plesiomorphic and which apomorphic must be done with caution."

The opinion of school 2 is clearly stated by Darlington (1970): "I think it can fairly be said that plants or animals that are not closely tied into stable, complex communities are ecologically more primitive than those

---

Vicariance and *vicariation* are defined and discussed on page 79.

that are thus tied. And it is in general the ecologically primitive, uncommitted forms that disperse and that populate new areas, while the ecologically derived-specialized forms are more likely to stay tied in their communities at their places of origin, again in exact contradiction to the Hennig-Brundin rule.''

There is, of course, no necessity to conclude that one or other school must be right and to attempt a judgment of the dispute on this basis, for they do not, between them, exhaust the possibilities.

Thus it cannot be ruled out that different kinds of organisms may be governed by different rules; that is, that some groups of organisms may "obey" school 1, others school 2. The obvious objection to this view is that it goes against the principle of parsimony in that it exemplifies, albeit in a very mild form, the antiscientific procedure of finding a special explanation for each special fact. But the alternative is to assume, perhaps mistakenly, that a large number of special facts all result from some single "law of nature," which we then strive to discover. This, too, may be a delusion. The frantic pursuit of general laws (whose discovery would make a person's reputation) is often, it seems to me, carried on at too low a level. For instance, in the present context, perhaps there is a law of nature just one step higher in the chain of cause and effect, a law explaining why some kinds of organisms behave one way and some another, instead of all alike. Or perhaps the search for generality will have to go much higher. The principle of parsimony seems often to be invoked as an excuse for what are in fact unjustifiably grandiose claims to generality.

Returning to the contrasting opinions of school 1 and school 2, on whether it is plesiomorphs or apomorphs that remain at their common center of origin, it should be noticed that there is actually no need to assume that any dispersal takes place at all. Suppose a chain of islands were occupied by different representatives of a group of related species and that the species formed a progression in the sense that each was more advanced (more apomorphous) than the one to the west of it (see Figure 3.4, bottom panel). Adherents of school 1 would conclude that the several species had evolved in the westernmost island where the most plesiomorphous descendant still remains, and that the more apomorphous descendants had dispersed eastward. Adherents of school 2 would conclude that evolution had taken place in the easternmost island, now occupied by the most apomorphous descendant, and that the progressively more plesiomorphous species had dispersed westward.

The observed distribution of species could come into being without any dispersal at all, however. As shown schematically in Figure 3.4, the observed sequence could as well result from the appearance of a succession of barriers separating the islands and preventing interbreeding be-

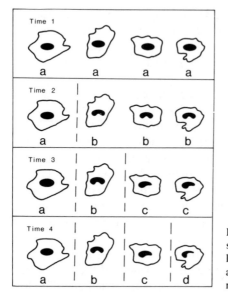

Figure 3.4 The development of successively more advanced taxa, in a row of islands, as a result of the formation one after another from west to east, of isolating barriers.

tween their populations. It may be argued that the appearance of isolating barriers one after another in a west-to-east row is improbable. Even so, it cannot be ruled out, and in considering and comparing competing explanations for an observed phenomenon, it is important to ensure that the list of possibilities examined is exhaustive. (The possibility illustrated diagrammatically in Figure 3.4 resembles the process postulated by proponents of school 3, as we see below.) Therefore the occurrence of a sequence (in the evolutionary sense) of taxa, in a sequence (in the spatial sense) of islands, as in the bottom panel of Figure 3.4, can in principle be explained in three different ways. There is thus no logical necessity to treat an observed sequence of this kind as an illustration of Hennig's "rule of progression" (Ashlock, 1974), which would treat school 1's interpretation as the only one possible.

An empirical example is due to Ashlock and Scudder (1966), who found that species of the lygaeid bug genus *Neocrompus*, occurring in a roughly east-west row of Pacific islands, formed an evolutionary "progression" as shown by the shape of their abdominal segments (see Figure 3.5). Ashlock (1974) considers that this is a case where Hennig's rule of progression does apply.

Very much stronger support for the opinion of school 1 is provided by cases where two or more "lines" diverge from a common center. Convincing examples are few. One, already mentioned in Chapter 2 (page 60), is that of the two sequences of Crustacean genera (family Parabathynel-

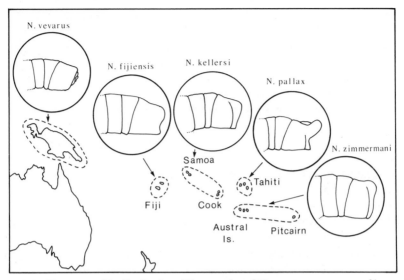

Figure 3.5 The geographic distribution of five species of *Neocrompus*, a genus of bugs of the family Lygaeidae (Heteroptera). The drawings show the terminal segments of the abdomen. The species form a progression from *N. vevarus* (the most primitive) to *N. pallax* (the most derived). (After Ashlock and Scudder, 1966.)

lidae) that appear to radiate in opposite directions from a common origin in eastern Asia (Schminke, 1974). Another concerns the distribution of families of Plecoptera (stoneflies) described by Illies (1965).

The "Vicariance Paradigm"

The proponents of school 3 call their theory the "vicariance paradigm."* Its gist is that, "On a global basis the general features of modern biotic distribution have been determined by subdivision of ancestral biotas in response to changing geography" (Croizat, Nelson, and Rosen, 1974). The splitting of populations that is a prerequisite for evolutionary divergence often results from the fragmentation of landmasses and the rafting apart of their biotas. Other causes of splitting are the formation of physiographic or climatic barriers such as mountain ranges or deserts. A

*Croizat and co-workers (1974) and Rosen (1975) use the term *vicariance* as a one-word synonym for *allopatric speciation*. It derives from the Latin "vicarius," a substitute. Numerous simultaneous speciations, all stemming from the division of one biota, constitute a *vicariant event* (Rosen, 1975) or *vicariation* (Brundin, 1967) and the words *vicariant* and *vicariant biota* connote the resulting new species or biotas. The set of words is most useful in providing concise terms for a set of related meanings.

salient point to notice is that it is the ranges of whole biotas, rather than of single species, that become fragmented. There are no centers of origin from which descendant species disperse; hence it is not necessary to invoke dispersal to explain observed allopatry of closely related species, though back-dispersal *after* speciation must be the cause of observed sympatry. To quote further: "Without a history of vicariance, the modern world biota would consist of only one or a few species, most if not all of which would be sympatric." And: "Vicariance, therefore, produces geographical differentiation and multiplication of species, and dispersal produces sympatry and the possibility of interspecific interaction (competitive exclusion, ecological differentiation, extinction)." To summarize: whole biotas are divided into geographically separated segments first, and evolutionary divergence, leading to speciation, comes second.

The theory disposes most satisfactorily of an unexplained paradox, the irreconcilability of allopatric speciation with the center of origin concept. But it does not explain how observed centers of origin, or at any rate, centers of high taxon density, are to be accounted for; and they are numerous (Figure 3.3 shows representative examples).

The theory explains convincingly the occurrence of whole "sister biotas," that is, matched sets of assorted sister species having a common within-set range, but with geographically separated ranges for the separate sets. Croizat and co-workers argue that it would be astonishing to find—as one does find—coincident geographic ranges in the Gondwana continents of such dissimilar kinds of organisms as freshwater fishes and crustaceans, crustacean parasites, earthworms, mollusks, birds, chironomid midges, stoneflies, mayflies, araucarias, *Nothofagus* species, proteas, and many other plant species, if speciation occurred first and dispersal second. The modes of dispersal of these taxa vary enormously. The only reasonable explanation is that separation precedes speciation.

Rosen (1975), on the basis of detailed studies, has shown how the pattern of modern geographic provinces in the Caribbean can be explained by supposing that an ancestral biota was fragmented, and the fragments then rafted, in accordance with the model shown in Figure 2.3. The model, which describes plate tectonic events in the Caribbean and Central American regions, stems from geophysical theory.

The arguments in favor of the vicariance paradigm seem to me so compelling as to be incontrovertible. But, as remarked above, they leave observed "centers of origin" unexplained. The reality of these species-rich "centers" is undeniable, and the problem of their existence cannot be ignored. Some speculative comments on the topic are offered in Sections 3 and 4.

3. SYMPATRIC, ALLOPATRIC, PARAPATRIC, AND QUASI-SYMPATRIC SPECIATION

In preceding sections it has been taken for granted that populations must be allopatric—geographically separated—if they are to undergo evolutionary divergence and become separate species. We now inquire whether observed biogeographic range patterns support this view. First it is necessary to define the terms that will be needed as clearly and unambiguously as possible, and in conformity with accepted usage. Perfection seems unattainable because of contradictions between (and even within) various authors.

The population unit whose fate we shall be considering is the *deme* (from the Greek "demos," people of a country). The term can only be applied to sexually reproducing species. According to most definitions a deme is an aggregate of individuals that, for at least one breeding season, interbreed panmictically; that is, all possible mating pairs have an equal chance of forming. Further, this aggregate is a spatially discrete breeding unit (Endler, 1977) whose members do not interbreed with those of other demes during a single breeding season. Notice that these two stipulations—complete panmixis within demes and no crossbreeding between them—amount to an overdefinition; one is left with no words to describe aggregates of organisms that possess only one of these attributes. Discussions of demes therefore contain the underlying assumption (usually tacit) that aggregates conforming to both specifications do, in fact, exist; and not only exist, but are the population units of which all sexually reproducing species are constituted. In what follows, we accept these assumptions without further comment in order to avoid intolerable prolixity.

Demes are assumed to be able to exchange genes between breeding seasons, through the exchange among them of dispersing individuals. After such dispersal, a newly migrated individual becomes one of the panmictically breeding members of the deme it has joined. Whether such exchanges between demes occur often or seldom depends on the spatial proximity of the demes and the vagility of the organisms concerned. Obviously the rate of gene flow between demes can vary from very rapid to zero.

Now consider how speciation by the splitting of an ancestral species actually takes place. It seems fair to say that nearly all students of evolution are agreed that (but for some exceptions noted below) the splitting takes place between demes. But there is considerable disagreement on the degree of isolation between neighboring demes that is necessary before they can diverge evolutionarily and found new species. Indeed, three different modes of speciation have been named that differ from each other in the degree of isolation envisaged. We consider them in turn.

The Different Modes of Speciation*

Sympatric speciation is defined (Mayr, 1970) as the origin of mechanisms leading to reproductive isolation within the dispersal area of the offspring of a single deme. It entails the differentiation of demes even though their members "are within cruising range of each other during the breeding season." If sympatric speciation occurs, it follows that "ecological isolates"—conspecific subpopulations with the same geographic range, distinguished by the fact that they prefer different local habitats—can be the forerunners of true species. It is not clear how reproductive isolation ever has a chance to develop since all individuals of the diverging demes are close enough to each other to interbreed. For this reason, few biologists regard sympatric speciation as possible, except among special kinds of organisms. Among animals, it is probably restricted to phytophagous and zoophagous parasites and parasitoids, most of which probably speciate in this way (Bush, 1975). Among plants, the appearance of allopolyploids (amphidiploids) is a form of sympatric speciation.

Allopatric speciation is the evolution of reproductive isolation following the separation of two or more subpopulations from each other by barriers that prevent the intermingling of their offspring. Stretches of ecologically unsuitable habitat, in which individuals of the species cannot maintain themselves, and across which they cannot disperse in one jump, constitute such barriers. This mode of speciation is regarded as the only possible one by those who suppose that demes cannot become intersterile unless gene exchange between them is totally prevented in this way. For speciation to occur, it is argued, gene exchange must be blocked for a sufficiently long time for the isolates to diverge to the point where they can no longer hybridize even if the opportunity presents itself again. Once speciation has occurred, the daughter species may become secondarily sympatric if the disappearance of barriers between them makes dispersal into each others' ranges possible again. Often they will have diverged ecologically, so that if sympatry is restored, the new species now occupy different habitats. This ecological differentiation is the result of speciation, not its cause.

*Their names embody the following Greek words: "sym," together; "allos," different; "para," by the side of; and the Latin "patria," native land.

Until it was realized that total cessation of gene flow between two demes is not a necessary precondition to the acquisition of reproductive isolation, most students of evolution regarded allopatric speciation as the most important, if not the only, mode of speciation (except in such special cases as those mentioned above). Recent studies of the way in which natural selection can counteract the effects of gene diffusion (see below) now make it impossible, in the words of Bush (1975), to accept the universality of allopatric speciation. So-called parapatric speciation may be as common as allopatric.

Parapatric speciation is defined as the evolutionary divergence of demes that are side by side, or contiguous. It is not a mere quibble to distinguish parapatric and allopatric speciation. True, both modes require that the differentiating populations should occupy nonoverlapping spatial regions; thus both are "nonsympatric." The distinction between them is not, however, that the gap between the differentiating demes is "negligible" in the case of parapatric, and "large" in the case of allopatric, speciation. The distinction is this: in allopatric speciation, the separated demes ultimately develop reproductive isolation *because* there is no gene flow between them. In parapatric speciation, the demes develop reproductive isolation *in spite of* a certain amount of gene flow between them. Theoretical modeling (Slatkin 1973, 1975; May, Endler, and McMurtrie, 1975; Endler, 1977) has shown that neighboring demes can achieve and maintain genetic distinctness even though they exchange genes, provided that the gene complement of a deme is selectively favored in its own region, and provided also that gene flow is not so great as to swamp the distinction maintained by selection (this topic is considered further in Chapter 10, page 305).

For speciation to go to completion, this distinction must, of course, increase to the level at which reproductive isolation is attained, whereupon gene exchange stops. Differentiation is hastened if heterozygotes are less fit than homozygotes. Although modeling studies have only considered one or a few gene loci, it seems reasonable to assume that the results are generally applicable. According to Woodruff (1978), "There is every indication that [gene] changes at a few loci are sufficient for speciation."

We now return to the question of how biogeographic patterns are affected by different modes of speciation. The patterns resulting from parapatric and allopatric speciation seem unlikely to differ appreciably, and in what follows the two modes are lumped together and termed "nonsympatric." At first thought, one would expect the ranges of species that had originated in this way to be disjunct or contiguous, or (if there has been post-speciation dispersal) to be overlapped. But one would not expect to find them nested one within another.

Such "nesting" is often observed however, and it need not imply that sympatric speciation has taken place. It is perfectly compatible with nonsympatric speciation if the spatial extent of demes is small. Probably this is often the case. Ehrlich and Raven (1969) offered evidence on a wide variety of plant and animal species showing that, for these species, each small, local population probably functions as a deme. In the light of their work it seems likely that nonsympatric speciation can take place in a much smaller geographic area than had previously been thought necessary. If this is so, the occurrence of "sister" species whose geographic ranges are nested need not necessarily mean that speciation has been sympatric. It may mean merely that a deme entirely surrounded by other demes has been able to differentiate nonsympatrically. The term *quasi-sympatric* speciation (Pielou, 1978b) can be used to describe the process. It connotes speciation that is actually allopatric or parapatric, but that appears to be sympatric because of the nested ranges of the resultant species. Figure 3.6 shows the process diagrammatically.

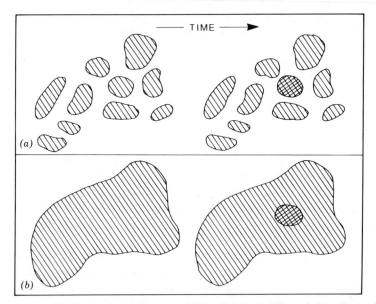

Figure 3.6 Quasi-sympatric speciation. (*a*) The evolutionary differentiation of one of several mutually isolated demes. (*b*) The same event, but misinterpreted; the separation of the species-population into demes has gone undetected and the differentiation of a central portion appears to be sympatric speciation.

The foregoing argument supposes that often what appears to be a single deme is, in fact, several separate demes; equivalently, that what appears to be a continuous geographic range is, in fact, a number of contiguous or narrowly disjunct subranges. If this be true, one is led to wonder why species are not *more* numerous than they are: why does not every deme develop into a distinct species?

The most probable answer is that stabilizing selection acts to prevent this. Thus we have two opposing forces at work. One, "deme fragmentation," favors the proliferation of new species. The other, stabilizing selection, favors the conservation of old species. What actually happens is the resultant of these two forces.

The effect of stabilizing selection in preventing what might be called overproliferation of species has been investigated by Dodson and Hallam (1977), who couch their discussion in terms of "catastrophe theory" (for a general account of the theory, see Zeeman, 1976). It is worth considering Dodson and Hallam's model in some detail.

A "Fold-Catastrophe" Model for Speciation

The model is illustrated diagrammatically in Figure 3.7. Consider a long geographic-environmental gradient. To take a particular example for clarity, consider a long stretch of north-south coastline. Suppose we are concerned with a species whose success is governed by some one environmental factor—mean annual temperature, say—that varies monotonically and continuously from one end of this coast to the other. The emphasis here is on slow continuous variation with no sudden steps, though the contrast between the environments at the two ends of the gradient is great.

The species of interest is assumed to be absent from the coast initially. It then invades from one end, say the south, and steadily expands its range northward. Assume that at all times the population as a whole consists of numerous small, reproductively isolated subpopulations, or equivalently that the species' range, from the southern end of the coast to the current position of the northward-moving front, is made up of a number of short, disjunct subranges. The gaps separating them are narrow but, nevertheless, act as genetic barriers.

Now suppose that at any latitude fitness is a function of phenotype, and that "phenotype" can be measured along one axis. This would be true if some one characteristic of the organism, its length for instance, were the only characteristic exhibiting phenotypic variation. (This assumption is made merely for clarity; it permits portrayal of a graph of fitness versus phenotype, at several different latitudes, in a space of three dimensions.

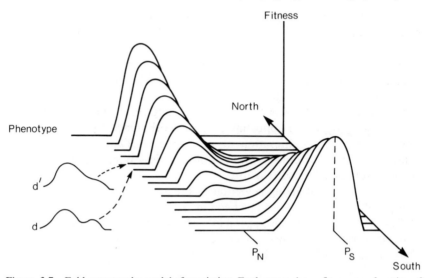

Figure 3.7 Fold-catastrophe model of speciation. Each curve shows fitness as a function of phenotype at a particular latitude. A species spreading north from a range initially confined to the far south will continue to have its southern phenotype, P_S, until there is no longer even a local maximum at P_S in the fitness curve. This point, somewhere between d and d', marks the boundary at which its phenotype switches abruptly to the northern form, P_N. (Adapted from Dodson and Hallam, 1977.)

The argument is perfectly general, however, and is not restricted to cases in which only one character varies.)

Denote by P_S the phenotype with maximum fitness at the southern end of the coast. As the species spreads northward, it will continue to have this phenotype even when the fitness versus phenotype curve becomes bimodal, as it does a short distance north of the starting point, and, indeed, even when the mode at P_S becomes lower than the other mode, at P_N (see Figure 3.7). This will happen as a result of stabilizing selection; that is, individuals whose phenotypes depart slightly from P_S will be of reduced fitness and will therefore be eliminated. At the same time, individuals whose phenotypes depart far enough from P_S to enjoy the greater fitness found in the neighborhood of the mode at P_N are unlikely to be produced. As a result, the P_S phenotype will be conserved. This will remain true until a point is reached along the coast (between d and d' in the figure) where the righthand ridge in the fitness surface stops being a ridge. At this point, a cross section through the surface, in the plane of the page, has an inflection rather than a secondary maximum at P_S. At this latitude, therefore, the phenotype will switch abruptly to P_N since selection can no longer stabilize the phenotype at P_S.

The salient feature of the model is that it explains why the series of isolated subpopulations differentiates into two species only, and why, where the changeover occurs, there is a sudden large switch in phenotype from P_S to P_N. Stabilizing selection prevents formation of a graduated series of species, each derived from one isolated subpopulation and each showing only slight phenotypic differentiation from its immediate neighbors. The sudden switch is a "catastrophe" in the terminology of catastrophe theory. In particular, it is a "fold-catastrophe," so named for a reason that is obvious from a glance at the fitness response surface.

Notice also that if the invader had colonized the northern end of the coast first and had then expanded its range southward (instead of vice versa as described above), the switch would not have occurred at the same latitude. It occurs at the point where the summit of the descending ridge—descending, that is, in the direction of migration—becomes an inflection. Thus given the fitness surface in Figure 3.7, the switchover latitude for a southward-advancing invader is south of that for a northward-advancing one.

This is not the only model that has been proposed to explain why there are not numerous species along a gradient, each derived from a small local subpopulation that has, by natural selection, evolved its own locally optimum phenotype or "adaptive peak." Wright (1967) suggested that a graduated series of local microspecies would indeed form first, but that a few superior ones would soon oust the rest; the successful few would reproduce to excess and their surplus members would diffuse into the ranges of their less successful neighbors. The latter would soon suffer competitive exclusion so that ultimately only a few well-differentiated species would remain.

Gradual and Abrupt Changes in Time

There is an obvious parallel between a one-dimensional space axis and the time axis. We have just discussed the changes that might occur in a species steadily expanding its range along a gradient in one dimension. Two contrasted outcomes were envisaged, either the formation of a series of many poorly differentiated species, or else the formation of a few well-differentiated ones. The latter outcome is made possible by stabilizing selection as illustrated pictorially by the fold-catastrophe model.

Now consider how a species changes through time. It can undergo a sequence of frequent slight changes; the result is a sequence of short-lived, poorly differentiated "microchronospecies." Or the changes can be much less frequent but, when they occur, more pronounced. The contrast can be likened to that between a staircase with many shallow steps and another with a few high steps. In evolution these contrasted processes can befall single lineages or the two divergent lineages that form when a parent species splits. In either case, they are usefully distinguished by the names "phyletic gradualism" and "rectangular speciation" (Stanley, 1975) respectively.

Mayr (1967) has emphasized that, since evolution goes on continuously and uninterruptedly, when abrupt speciation takes place it must alternate with another form of evolution. He calls the two forms "maintenance evolution," the stabilizing selection that prevents an adapted species from drifting away from the optimum it has achieved; and "switch evolution," the sudden jump from one local fitness peak (or locally optimum phenotype) to another. In terms of the staircase analogy, maintenance evolution represents the treads and switch evolution the risers.

It is worth reemphasizing that natural selection goes on unceasingly. It occurs both within and between species. Stanley (1975) calls these two modes "microevolution" and "macroevolution." Microevolution is within-species evolution and results from differential survival of individual organisms. Macroevolution is the formation of new species and results from differential survival of whole populations. There is no necessary connection between the rates of these two processes; in Stanley's words, they are decoupled.

To summarize these various terminologies: macro- and microevolution refer to speciation and within-species evolution respectively; phyletic gradualism and rectangular speciation are alternative possible modes of speciation; and in a lineage that undergoes rectangular speciation, two different processes, maintenance evolution and switch evolution, must alternate in time.

The view that rectangular evolution is a commoner form of macroevolution than phyletic gradualism is gaining ground among paleontologists. There are two chief reasons for this.

On the one hand, rectangular speciation explains why graded sequences of fossil forms are rarely found, even in species for which there is an unbroken fossil record of long duration (Eldredge, 1971). It is true that lineages showing slow, directional character trends are known, for example, in the foraminifera (Scott, 1976), but they are fairly uncommon. It therefore seems probable that so-called gaps in the fossil record are often not gaps at all but merely mark an abrupt change in phenotype in an uninterrupted record. If phyletic gradualism is uncommon, there is no reason to expect to find a series of missing links between one species and its conspicuously different descendant. Such missing links probably never existed and failure to find them does not indicate a stratigraphic gap (Eldredge and Gould, 1972).

The other reason for supposing that phyletic gradualism seldom or never operates is that there seems never to be any correlation between a lineage's rate of evolution and the generation time of the organism concerned. If evolution were gradual, one would obviously expect change to be faster in species whose generation times were short. Such a relationship has not been found.

Stanley (1975) discusses speciation rates in various groups. If we assume that, within some chosen taxonomic group, ancestral species are splitting and some of the resultant isolates are diverging to form new species, then the total number of species is growing exponentially. Let r denote the instantaneous rate of increase in species number, that is, the change in number of extant species, per existing species, per unit time. (Notice that r is the difference between the rate of formation of new isolates and the rate at which lineages go extinct. It may be positive, zero, or negative. To say that the number of species in a group is growing exponentially does not automatically imply, though it does not rule out, "explosive" increase.)

Next, let t denote the age of a taxon, as judged from the fossil record, and N the number of species of the taxon now living. Assume that all these species are descended from a single ancestral species. Then r may be obtained from the relation

$$N = e^{rt},$$

whence

$$r = \frac{(\ln N)}{t}$$

The following are two numerical examples given by Stanley (1975).

For the slowly evolving pelecypod family Tellinidae, $N \simeq 2700$ and $t \simeq 120$ m.y., whence $r \simeq 0.07$ species per species per million years.

For the mammalian family Bovidae (cattle, sheep, goats), $N \simeq 115$ and $t \simeq 23$ m.y., whence $r \simeq 0.21$ species per species per million years.

Numerical comparisons such as these should not be accepted uncritically, however. They may do more than prove that morphologically complex organisms give the appearance of evolving faster than simple ones merely because their structural intricacy allows them to display a greater variety of visible changes. It does not automatically follow that invisible changes, particularly evolutionary changes in the underlying genomes, occur more slowly in "simple" organisms. Arguments for and against this view, and its implications, have been marshalled by Eldredge and Gould (1976), though an acceptable method for testing it has yet to be proposed.

4. THE LATITUDINAL DIVERSITY GRADIENT

We now return to the problem of whether it is possible to reconcile the vicariance paradigm with the frequent occurrence of what may be called "centers of high taxon density," even if not "centers of origin"; the latter name begs a question of course.

The reality of these centers of high taxon density seems undeniable. Figure 3.3 shows three out of a very large number of known examples. In what follows, they are called *diversity-islands* for brevity. Thus Figure 3.3*a* shows the genus diversity-island for the Phasianinae, and Figure 3.3*b* shows the species diversity-islands for the Crocodilia.

The great majority of diversity-islands have their centers in the tropics and their long axes parallel with the parallels of latitude. Consequently, when the species or genus diversity-islands for a large number of families or orders (or more inclusive taxa) are considered together, or equivalently, when their isopleth maps are superimposed, it is seen that biotic richness in general attains its greatest value in the equatorial region and falls off steadily with increasing latitude to north and south. That is, there is a latitudinal diversity gradient.

Thus the two following statements amount to two ways of saying the same thing, and one explanation suffices for both.

1. Diversity-islands exist in great numbers and the majority have their centers at low latitudes.
2. Overall biotic diversity exhibits a latitudinal gradient, decreasing from the equator toward both poles.

The search for an explanation can concentrate on proximate or ultimate causes. The best way to proceed seems to me to consist in hypothesizing proximate causes of a theoretical kind—that is, models—and deferring for later consideration the search for an ultimate cause, that is, for the reason some particular model or combination of models seems the most likely proximate cause.

Two conceptually possible models can reasonably be discarded right at the outset since, on the basis of existing knowledge, their probability is so low it is not worth lingering over them. These are that sympatric speciation is common; or else that new species arise with equal frequency in all latitudes and then migrate to the tropics.

We therefore have to explain why nonsympatric (or quasi-sympatric) speciation is much more prolific in some parts of the world than others. Two explanatory hypotheses, one static, the other dynamic, are worth contemplating. We consider them in turn and then show that, from a practical point of view, they are the same.

The Static Model.   Two factors, possibly independent, foster multiplication of species. These are, first, the fragmentation of a parent species-population into a large number of mutually isolated demes; the more fine-grained the fragmentation pattern, the greater the number of potential new species. The second factor is the existence of many distinct locally optimum phenotypes (adaptive peaks*); the more numerous these peaks (which stabilizing selection tends to conserve by eliminating "stragglers"), the greater the number of potential new species.

In any region, therefore, two factors control the number of potential species. It will clarify the discussion to show them in a 2 × 2 table. The capital letters in the cells of the table label the four contrasted states to be envisaged. They are considered below.

| | | Adaptive Peaks | |
		Few	Many
Fragmentation into demes	Coarse	A	B
	Fine	C	D

The consequences to be expected given these four combinations of factors are:

A: The region will contain only a few species. They will be well differentiated and will occupy extensive geographic ranges.

B: As for A. The demes are few (because fragmentation is coarse) and the number of species cannot exceed the number of demes. Thus

---

*If fitness is plotted against phenotype and there are many, say n, phenotypic variables (rather than just one, as in Figure 3.7), then the graph occupies a space of $n + 1$ dimensions. The set of all points showing the fitness associated with every conceivable phenotype is a hypersurface. Peaks in this hypersurface are *adaptive peaks*. Each occurs "over" a point in n-dimensional "phenotype space" corresponding to a locally optimum phenotype. It is important to observe that the discussion in this note refers to abstract, conceptual space, *not* to space on the ground. Additional axes would be required if we wished to show how a fitness hypersurface differed from one place to another.

only a subset of the many possible well-adapted phenotypes will become differentiated as species; which ones these are is a matter of chance; they will be conserved by stabilizing selection.

C: The number of well-differentiated species cannot exceed the number of adaptive peaks. Hence, if there are more demes than peaks, not all demes can evolve into well-differentiated species. Some peaks are of necessity "shared" by a group of separate demes, and all demes in such a group have closely similar, or identical, phenotypes. Although, in some cases, the separate demes may lack the opportunity to hybridize, they may or may not lose the capability to hybridize. If they do lose it, that is, if they become intersterile, then they become true species, albeit "cryptospecies" (morphologically indistinguishable species).

D: The region will contain many true, distinct species.

According to this model, therefore, a taxon will be rich in species in a given region provided it is fragmented into numerous isolated demes and provided, also, there are numerous adaptive peaks.

The Dynamic Model.   This model assumes that the logistic law of population growth is applicable also to the proliferation of species. The purpose of contemplating the model, however, is not to make predictions, but merely to translate some elementary concepts into easily manipulable symbols. As is well known, the so-called saturation (equilibrium) level, say K, of a population whose growth is governed by the logistic law, may be written as a fraction, say $K = r/s$, where r denotes the intrinsic rate of natural increase, and s is a parameter measuring the intensity of competition among individuals (Pielou, 1977).

Now suppose the same law governs the growth of the "population" of species belonging to some higher taxon, and let the same symbols be retained unaltered because of their familiarity. Now r denotes the rate at which species-populations split, or demes proliferate, and s is a parameter measuring the intensity of interspecies competition. If K, the number of species of a given taxon, is larger in one region than in another, it must follow that r is larger or s smaller (or both) in the species-rich region. But to say that r is large is tantamount to saying that an ancestral population is fragmented into numerous demes. And to say that s is small is tantamount to saying that the number of phenotypes that can coexist without any being excluded by competition—that is, the number of adaptive peaks—is large.

We now have two interchangeable models, either of which serves as a proximate "explanation" of observed differences in species-richness. They are not so much explanations, of course, as alternative ways of framing the question with which we began, namely why is species-richness so much greater in some parts of the world than in others? For terrestrial communities, species-richness tends to be greatest in low latitudes, especially where the climate is moist as well as warm. In such regions, biotic productivity is at its highest, and is continuous (non-seasonal). It therefore remains to decide whether, and if so, why (in the terminology of the static model) these conditions are conducive of a high degree of fragmentation of species-populations, and also of a large number of adaptive peaks (equivalently, the possibility of unhampered coexistence of a large number of species). The answers, whatever they are, will be the ultimate explanation of high diversity.

There is no shortage of plausible arguments purporting to explain why these conditions should lead to high diversity. The number of attractive theories is, indeed, more than sufficient. Many have been summarized by Pielou (1975). But the constant repetition of what were once speculative suggestions has gradually converted them into "self-evident truths" without their having been adequately tested. As a result, the cause of the latitudinal diversity gradient is obvious to some and still a mystery to others, depending on their preconceptions.

It is particularly unilluminating to say that ecological "niches" are numerous where diversity is high. This is merely a way of rephrasing, not of answering the question, and an ambiguous way at that. This is because the word "niche" itself is ambiguous.

Some ecologists use the word to mean the "Grinnell niche" defined as the totality of sites in a region having given environmental conditions (Grinnell, 1917). To say that Grinnell niches are numerous is very roughly equivalent to saying that a region is fragmented into a large number of varied habitat patches and, by extension, that species are fragmented into a large number of isolated demes.

To other ecologists, a niche is a "Hutchinson niche," the set of all conditions needed to ensure the successful survival and reproduction of a species (Hutchinson, 1957). To say that Hutchinson niches are numerous is equivalent, again very roughtly, to saying that adaptive peaks are numerous in phenotype-space.

In brief, the Grinnell niche is actual, the Hutchinson niche, conceptual. The Grinnell niche consists of real places; the Hutchinson niche is an abstract hypervolume in a graph of many dimensions.

In searching for the factors governing diversity, it is important to keep separate what are probably the two most decisive contributory causes: a high degree of population fragmentation, and a large number of adaptive peaks or viable phenotypes. These factors are confounded when explanations are offered that use the word "niche"* without defining it.

In the opinion of many biogeographers, the search for an ultimate cause for the latitudinal diversity gradient has not so far yielded satisfactory results, in spite of the effort that has been devoted to it. The time has come, I think, to look for entirely new angles of attack. The continual refinement of theories with a strong element of circular argument is leading nowhere. To say that where conditions are "favorable to life" life is favored, and hence that species multiply, leaves unanswered the really basic question, which is: why, in the course of organic evolution, have fewer species become adapted to high latitude than to low latitude conditions? It takes no great effort of the imagination to visualize a fauna for which the conditions we call "benign" are "harsh" and vice versa: Thinking of the tropics as benign and the polar regions as harsh is only a habit of thought; it *results* from the fact that life is more abundant in the tropics and our own species originated there.

Further progress on this fascinating problem is unlikely to come until uncritical preconceptions are abandoned.

Changes in Diversity through Geological Time

Knowledge about the way in which the worldwide total of organic species—the diversity of the whole biosphere—has changed through geological time would be extraordinarily interesting if it could be acquired. But the evidence on which conclusions might be based is sketchy and hard to interpret.

There are two opposing views on the topic (Gould, 1976). One is that diversity has steadily increased with time; the other, that an equilibrium level of diversity was reached early in the history of the earth.

According to the first view, the evolution of the biosphere is, and continues to be, directional, the trend being towards ever greater morphological complexity in organisms of all kinds. The greater the complexity of individual organisms, the more scope there is for diversification, that is, the appearance of many different kinds of organisms.

According to the second view, diversity reached an equilibrium level—in other words, the number of species reached saturation—in the distant past, perhaps as early as the Cambrian, 600 m.y. ago. Since then, although the kinds of organisms living at any one time have been remark-

---

*There is yet a third kind of niche, the "Elton niche" (Elton, 1927), defined as the function a species performs in a community. In some respects this concept combines the other two.

ably different, the most recent change for example being that from the Mesozoic "age of reptiles" to the Cenozoic "age of mammals," the total number of species has fluctuated around a constant value. This saturation level is a property of the biosphere itself; the fluctuations are caused by changing physical conditions, that is, by extrinsic changes.

As an example of such a fluctuation, Gould (1976) gives the wave of extinctions that is believed to have greatly reduced the diversity of the marine fauna of the continental shelves at the time of the Permian-Triassic transition. The extrinsic cause he adduces is, very briefly, as follows. At the end of the Paleozoic era, when all the world's continents became united to form Pangaea, the crustal plates were, for a time, motionless. Since they were not being forced apart by widening rifts, the formation of new oceanic crust to fill these rifts stopped also. But newly formed crust always solidifies first in the form of midocean ridges. Therefore, when the plates stopped moving, the ridges collapsed. This, in turn, meant the disappearance of enormous masses of rock that had been displacing sea water. Consequently the sea level fell, converting to dry land much of what had been submerged continental shelf. The final effect was a great reduction in number of individuals, and hence of taxa, of shallow-water marine forms because the areal extent of their environment was so greatly reduced.

This long chain of cause and effect is one possible explanation of a reduction below saturation level of an important component of the biosphere, the marine shelf fauna. An alternative, or contributory, explanation of the drop in species-richness is that the formation of Pangaea from previously separate continents entailed the formation around its margin of one long, unbroken continental shelf. What had been separate segments of shelf were united into one long shelf, and extinctions followed as a result of competitive exclusion; as well, the shelf environment was comparatively uniform (Valentine and Moores, 1972). The exclusion process was similar to that adduced by Kurtén (1969) to explain the extinction of many mammalian orders in the late Tertiary (see page 56).

As with the explanations for the present-day latitudinal diversity gradient, there is no shortage of plausible theories about what happened in the past; others are described by Pielou, 1975. Moreover there is great uncertainty as to what did happen, and hence what the events were that need to be explained. Raup (1972) has emphasized how easy it is to be misled by systematic biases in the fossil data available for interpretation.

He points out, for example, that the apparent decrease in diversity of shallow-water marine forms in Permian-Triassic time (for which two explanations were offered above) may simply stem from the fact that not much sediment was laid down at that time. Opportunities for the preservation of fossils were therefore poor and the available samples of contemporary life are rather small.

This leads to bias when one attempts to infer the diversity of different hierarchical levels—species, genera, families, orders, and the like—that were present in the population sampled. A sample of given size necessarily contains a larger proportion of the total (population) number of orders than of families, a larger proportion of families than of genera, and of genera than of species. Consequently a particular order of, say, the Pelecypoda (bivalve mollusks) will seem to be less diversified—in the sense of being represented by fewer families—in a small fossil assemblage than in a large one. Thus the diversity within each level of the taxonomic hierarchy is strongly influenced by the quantity of fossiliferous sediment available for study, and this varies from one epoch to another because of physical factors.

*5. THE GEOGRAPHIC RANGES OF RELATED SPECIES

It was remarked on page 84 that closely related species are often found to have ranges nested within each other. A mechanism that might account for this, without invoking sympatric speciation as an explanation, was described; it is illustrated in Figure 3.6. This exemplifies the fact that it is not difficult to devise models leading from any given speciation mode to any chosen biogeographic pattern, and it may well be that the study of biogeographic patterns can offer little help toward discerning modes of speciation in particular cases. However, a statistical study of biogeographic patterns and the way they overlap is worthwhile. If "overlap patterns" can be characterized quantitatively, and if comparisons can be made among empirical frequency distributions of the measurements obtained from different regions and from different taxonomic groups, the results may well prove illuminating in ways that cannot yet be foreseen. A way of tackling the problem is described in the first subsection below. It treats species' geographic ranges "coarsely"; that is, no distinction is made between continuous ranges and interrupted, or "gappy," ranges. Such a distinction is usually impossible because of the inadequacy of available data. For lack of detailed information, one must usually treat a species' range as roughly coextensive with the convex hull surrounding points on the map where the species was found (but see page 279).

To treat ranges thus, while useful in some investigations, is apt to lead to neglect of the fact that ranges often are "gappy." Gaps are more likely the rule than the exception; it was argued above (page 81) that the majority of species-populations are probably made up of demes, and these may be narrowly or widely disjunct. However, inadequate sampling masks the gaps among them, and may lead one to infer that a range is far more continuous than in fact it is. This is the topic of the second subsection below.

The Statistics of "Sheaves" of One-Dimensional Ranges

A map showing the ranges of several species can exhibit many different patterns; the object of a statistical study of such patterns is to compare the observed frequencies of the various forms with the frequencies predicted by theoretical models. Also, one can intercompare observed frequency distributions obtained from different regions, or relating to different categories of organisms.

First consider the patterns that the geographic ranges of a pair of species can present. There are three possibilities: the two species can be disjunct; they can "lap" each other; or they can be nested; see Figure 3.8, from which the definitions of these terms will be clear. Inferring the actual patterns of species' ranges from records in the literature is much easier to do for littoral species, whose ranges are representable as line segments, than for species whose ranges are irregular two-dimensional shapes. Therefore the following discussion is limited to species with "linear" ranges. The three possible patterns for a pair of such species are shown in the lower panel of Figure 3.8.

Now consider how the "overlap" of a pair of species may be measured. One may use a bivariate measure or a univariate measure.

The bivariate measure, (m, q), is defined as

$$(m, q) = \begin{cases} (0, 0) & \text{if the pair is disjunct,} \\ (1, 0) & \text{if the pair is lapped,} \\ (0, 1) & \text{if the pair is nested.} \end{cases}$$

The univariate measure, λ, is defined as

$$\lambda = m + 2q.$$

Therefore $\lambda = 0, 1$, or 2 depending on whether the pair is disjunct, lapped, or nested.

To measure the overlap of the "sheaf" of ranges of a group (for

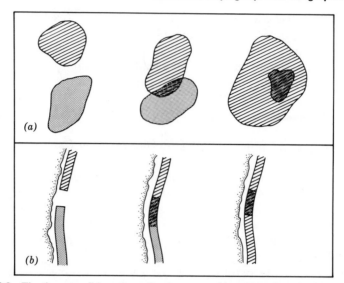

Figure 3.8 The three possible patterns for the geographic ranges of a pair of species. They may be disjunct (as on the left), "lapped" (center), or "nested" (right). (*a*) The patterns in a two-dimensional region; (*b*) as they would be exhibited by littoral species whose ranges can be treated as one-dimensional.

example, a genus) of s related species with $s > 2$, we may again use a bivariate or a univariate measure. We define the bivariate measure as

$$(M_s, Q_s) = \sum_{\substack{\text{all} \\ \text{pairs}}} (m, q),$$

and the univariate measure as

$$L_s = \sum_{\substack{\text{all} \\ \text{pairs}}} \lambda.$$

The summations are over all $s(s - 1)/2$ possible pairs of the s species. The symbols m, q, and λ relate to a species-pair that forms a subset of the species in a genus; M_s, Q_s, and L_s relate to an s-species genus or other s-member related group. For a two-species genus, the appropriate symbols are M_2, Q_2, and L_2; the "summation" is over a single term so that $L_2 = \Sigma\lambda = \lambda$, and similarly for M_2 and Q_2.

Figure 3.9 demonstrates the measurement of (M_s, Q_s) and L_s for a sheaf of $s = 5$ ranges. Observe that the method of measurement takes no

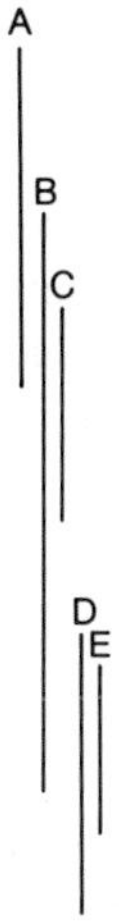

Pair	m	q	λ
A B	1	0	1
A C	1	0	1
A D	0	0	0
A E	0	0	0
B C	0	1	2
B D	1	0	1
B E	1	0	1
C D	0	0	0
C E	0	0	0
D E	0	1	2
	4	2	8

$$(M_5, Q_5) = (4, 2)$$

$$L_5 = M_5 + 2Q_5 = 8$$

Figure 3.9 Finding the overlap of a sheaf of littoral species. The ranges are aligned along a "north-south" shoreline. There are $s = 5$ species and hence $s(s - 1)/2 = 10$ species-pairs. The table shows m, q, and λ for each pair, and their sums, M_5, Q_5, and L_5.

account of the lengths of the overlaps. Thus if we write A_1 and A_2 for the northern and southern limits (endpoints) of species A's range and correspondingly for species B, and if the limits occur in the order $A_1 B_1 A_2 B_2$, say, then the overlap of the pair is $(m, q) = (1, 0)$, λ = 1, regardless of the length of shoreline between B_1 and A_2.

We now inquire: what are the relative frequencies of disjunct, lapped, and nested species, given some suitable null hypothesis, and what is a suitable null hypothesis? These problems have been discussed by Pielou (1977a,b, 1978a,b).

Two quite different null hypotheses can be entertained and each has a claim to be thought "natural." Both embody the assumption that the ranges of related species are located independently of each other. They differ in that the first does not take the lengths of the ranges as given, whereas the second does. In describing them it is assumed, for ease of exposition, that the shore along which ranges are aligned trends north-south. The hypotheses concern the overlap pattern of a "sheaf" of s ranges.

The first hypothesis (H1) is that the northern and southern limits of the s ranges are randomly permuted subject only to the obvious constraint that a species' northern limit must lie north of its southern limit.

The second hypothesis (H2) assumes that the s ranges have their observed geographic lengths (measured in kilometers, for example). Then it is hypothesized that these ranges are located independently and at random in a one-dimensional space of given length, namely the observed range of the whole genus. In concrete terms, one envisages a long, narrow box (the range of the genus); then s sticks of appropriate lengths (the ranges of the species) are put independently and at random into the box.

The biogeographic preconceptions underlying the two hypotheses are entirely different. To see this, consider a two-species genus whose species, A and B, are observed to have nested ranges. That is, their northern and southern limits occur in the sequence $A_1B_1B_2A_2$. The range of the genus coincides, by definition, with that of species A; the range of species B is observed to be shorter than that of species A.

Then, under H2, it is clear* that

$$\Pr\{(M_2, Q_2) = (0, 1) \mid H2\} = \Pr\{L_2 = 2 \mid H2\} = 1.$$

In words, nesting is inevitable given the lengths of the two ranges.

However, under H1 all that is given is the existence of two species. The three possible sequences of their range limits, namely $A_1A_2B_1B_2$, $A_1B_1A_2B_2$, and $A_1B_1B_2A_2$, are treated as equiprobable so that

$$\Pr\{(M_2, Q_2) = (0, 1) \mid H1\} = \Pr\{L_2 = 2 \mid H1\} = \tfrac{1}{3}.$$

Whether to entertain H1 or H2 thus depends on the answer to the question: Is the length of a littoral species' range an innate property of the species, an attribute the species would possess regardless of the existence of other species? If one thinks the answer is "yes," H2 is the hypothesis to adopt; if "no," H1. It is, in any case, a matter of opinion, biased (perhaps) by the fact that H1 is far more tractable, mathematically, than H2.

We now explore hypothesis H1. Notice first that L_s can take the values $0, 1, \ldots, s(s - 1)$; M_s and Q_s can take the values $0, 1, \ldots, s(s - 1)/2$. Further, since a pair of species cannot be simultaneously lapped and nested [equivalently, since $(m, q) = (1, 1)$ is impossible], we must have $M_s + Q_s \leq s(s - 1)/2$. The equality holds only when every species-pair is nondisjunct, that is, is either lapped or nested.

Obviously, when all s species are mutually disjunct, $(M_s, Q_s) = (0, 0)$ and $L_s = 0$. When every species-pair is lapped (as are the cards in a slightly skewed deck of cards), $(M_s, Q_s) = (s(s - 1)/2, 0)$ and $L_s = s(s - 1)/2$. When each species' range is nested within that of the next larger range, $(M_s, Q_s) = (0, s(s - 1)/2)$ and $L_s = s(s - 1)$. (See Figure 3.10.)

---

*The symbolic expression $\Pr\{L_2 = 2 \mid H2\}$, for example, means "the probability that $L_2 = 2$ given that hypothesis H2 is true."

Some of the consequences of hypothesis H1 are the following (for conciseness, the subscript s has been dropped):

1. The distribution of L is symmetrical. That is, $\Pr\{L\} = \Pr\{s(s-1) - L\}$.
2. The expected value of L is $E(L) = s(s-1)/2$. This follows from 1.
3. From 1 and 2 it follows that $\Pr\{L < E(L)\} = \Pr\{L > E(L)\}$.
4. The bivariate distribution of (M, Q) is symmetrical about $M = Q$.

That is, $\Pr\{(M, Q)\} = \Pr\{(Q, M)\}$; and $\Pr\{Q < M\} = \Pr\{Q > M\}$.

Suppose now that we are interested in the overlap patterns of the species within each genus for a large collection of genera in a family, order, or higher ranking taxon of littoral organisms. If all the necessary data are available, we can determine (M, Q) and L for each genus. Each numerical value depends on s, of course, and s varies from genus to genus. Therefore values from genera with different numbers of species are not comparable with each other. Consequently, we cannot combine the observations from all genera to obtain a single empirical frequency distribution of (M, Q) to compare with expectation.

However, two questions can be asked concerning each genus: Does L exceed or fall short of expectation? And does Q exceed or fall short of M? The answers enable one to assign every genus to one of four categories or groups regardless of its number of species. The groups are:

Group 1: Genera with $L > E(L)$ and $Q > M$;
Group 2: Genera with $L < E(L)$ and $Q < M$;
Group 3: Genera with $L > E(L)$ and $Q < M$;
Group 4: Genera with $L < E(L)$ and $Q > M$.

(*NOTE:* Genera having $L = E(L)$ or $M = Q$ are shared between the appropriate groups.)

The probability, under H1, that a genus will be found to belong to any of these groups varies with s, but the variation is very small and the probabilities rapidly approach limiting values as s increases (Pielou, 1978b).

Thus, writing p_i for the probability that a genus belongs to group i, for $i = 1, 2, 3, 4$, it is found that $p_1 = p_2$ lies in the range $(0.30, 0.33)$ and tends slowly to 0.30 as s increases. (The numerical values are correct to two decimal places.) Likewise, $p_3 = p_4 = 0.5 - p_2$ lies in the range $(0.17, 0.20)$ and tends to 0.20. For practical purposes, therefore, we may treat the p_i as independent of s and write $p_1 \simeq p_2 \simeq 0.3$ and $p_3 \simeq p_4 \simeq 0.2$ for all s ($\Sigma p_i = 1$).

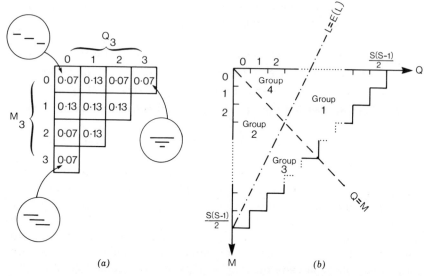

(a) M (b)

Figure 3.10 (a) The bivariate probability distribution of (M_3, Q_3) under hypothesis H1. The fraction in each cell of the table is the probability that, given the hypothesis, the sheaf of ranges of a 3-species genus will have the (M_3, Q_3) value specified by the cell's row and column headings. The circled diagrams at each corner show the three "extreme" sheaf patterns. (b) The general form of the (M, Q) table for any s, divided into the four groups. The dashed line $Q = M$ divides the space into upper and lower halves in which $Q > M$ and $Q < M$ respectively. The dash-dot line $L = E(L)$ [equivalently, $M + 2Q = s(s - 1)/2$] divides the space into left and right halves in which $L < E(L)$ and $L > E(L)$ respectively.

Notice that $p_1 + p_4 = 1 - (p_2 + p_3) = 0.5 = \Pr\{Q > M\} + \frac{1}{2}\Pr\{Q = M\}$. Similarly, $p_1 + p_3 = 1 - (p_2 + p_4) = \Pr\{L > E(L)\} + \frac{1}{2}\Pr\{L = E(L)\}$.

Since the genera have been classified on the basis of two attributes, the results can be tabulated in a 2×2 table; [recall that genera having $L = E(L)$ are equally divided between the classes with $L < E(L)$ and $L > E(L)$; and likewise for genera with $M = Q$]:

	$L > E(L)$	$L < E(L)$
$Q > M$	Group 1 $p_1 \simeq 0.3$	Group 4 $p_4 \simeq 0.2$
$Q < M$	Group 3 $p_3 \simeq 0.2$	Group 2 $p_2 \simeq 0.3$

The positions of these groups in the probability space of (M, Q) are shown in Figure 3.10b.

The salient result is this: the variations of the p_i values with s are so negligibly small that data on all the genera (or other species groups) in a large class of organisms can be combined even though their numbers of members vary. This greatly facilitates comparisons.

For example, Pielou (1978b) compared the overlap patterns within genera of seaweed species on the Pacific and Atlantic shores of North and South America. Two results of biogeographic import emerged.

The observed values of $p_1 + p_4$ are roughly equal on the two shores and significantly exceed their expectation under H1, namely 0.5. That is, the geographic ranges of congeneric species tend to be nested much more often than hypothesis H1 predicts. This suggests that speciation among these algae may be sympatric or quasi-sympatric (see page 84).

Comparing the Atlantic and Pacific shores, it is found that the observed value of $p_1 + p_3$ is significantly greater on the Atlantic shore than on the Pacific. That is, the geographic ranges of congeneric species overlap one another to a greater extent on the Atlantic; or, put the other way, within-genus disjunctions are relatively more common on the Pacific. Possible explanations for this result are discussed in Chapter 5 (page 164).

The Extent and Number of Demes
Comprising a Population

It was remarked on page 85 that breaks in the continuity of biogeographic ranges often go undetected. Unless a species is very well known, or occurs only in regions very well studied, its range may contain unsuspected gaps that have the effect of subdividing it into subpopulations (demes) between which genes are exchanged only occasionally or not at all.

This subsection explores the effect of inadequate sampling on inferences about ranges. The discussion takes for granted a knowledge of simple Markov chain theory. A very clear, elementary introduction to the subject is contained in Howard (1960). We shall again be concerned with species whose ranges are effectively one-dimensional and can therefore be represented on a map by line segments (not necessarily straight). Aquatic organisms living in rivers, or littoral organisms living along shores, are examples. For brevity, the region in which the species of interest occurs will be called a shore.

Suppose we wish to infer the number of demes into which the population of a littoral species inhabiting a given shore is subdivided, and the mean lengths of the demes and the gaps between them. Assume that we know the minimum distance that must separate two demes for them to maintain their discreteness. Take this distance as the unit of length. Any break in continuity larger than this distance we define as a "gap"; segments of the population on either side of a gap necessarily belong to different demes.

If the shore is examined for presence or absence of the species at a row of sampling stations evenly spaced at one unit intervals, we can acquire "complete" knowledge about the species' range and its subdivisions. No gaps will go undiscovered. A gap consists of a sequence, or run, of one or more empty stations and the "run-length" of the gap is defined as the number of empty stations forming it. Likewise, a run of occupied stations is assumed to coincide with a deme. Observe that distance is being treated as a discrete variate; whenever the species is found at two adjacent sampling stations, then, *by definition*, there is no gap in its range at that point.

Next assume the sequence of occupied and empty stations along the shore constitutes a simple two-state Markov chain. An occupied station is in state 1; an empty station is in state 2. The Markov matrix $\mathbf{P}$ is

$$\mathbf{P} = \begin{pmatrix} p_{11} & p_{12} \\ p_{21} & p_{22} \end{pmatrix}$$

where p_{ij} is the probability that, if the stations are examined in sequence, a station in state i will be succeeded by a station in state j for $i, j = 1, 2$. Since there are only two states, we must have $p_{i1} + p_{i2} = 1$ for $i = 1, 2$. Because of this, it is convenient to use unsubscripted symbols for the probabilities and write

$$\mathbf{P} = \begin{pmatrix} 1 - \lambda & \lambda \\ \mu & 1 - \mu \end{pmatrix}.$$

It is assumed that $\lambda, \mu \leqslant 0.5$; equivalently, that a "step" from one station to the next is more likely than not to bring no change of state.

Now write d_1 for the length of a deme and g_1 for the length of a gap. Their expected (mean) values are (from Markov chain theory)

$$E(d_1) = \frac{1}{p_{12}} = \frac{1}{\lambda} \quad \text{and} \quad E(g_1) = \frac{1}{p_{21}} = \frac{1}{\mu}.$$

We have assumed so far that the distance between sampling stations was one unit. Observations spaced as closely as this automatically give complete information, free of error, on the lengths of the alternating demes and gaps.

Now suppose such detailed sampling is impracticable and it is possible to examine only every nth station ($n > 1$). The observed sequence of occupied (state 1) and empty (state 2) stations will again form a simple Markov chain, but this time with Markov matrix $\mathbf{P'}$, say, where

$$\mathbf{P'} = \mathbf{P^n} = \begin{pmatrix} p'_{11} & p'_{12} \\ p'_{21} & p'_{22} \end{pmatrix}.$$

The mean run-lengths (numbers of stations) of the demes and the gaps, say $E(d_n)$ and $E(g_n)$, are now

$$E(d_n) = \frac{1}{p'_{12}} \quad \text{and} \quad E(g_n) = \frac{1}{p'_{21}}.$$

However, because the stations are now spaced at intervals n times as great as before, the unit of length is n times as great. In terms of the original units, the mean lengths of the demes and gaps are thus $nE(d_n)$ and $nE(g_n)$.

It can be shown that

$$\mathbf{P'} = \mathbf{P^n} = \frac{1}{1 - \delta} \begin{pmatrix} \mu + \lambda\delta^n & \lambda - \lambda\delta^n \\ \mu - \mu\delta^n & \lambda + \mu\delta^n \end{pmatrix}$$

where $\delta = 1 - (\lambda + \mu)$.

Therefore, the apparent mean lengths, in original units, of the demes and gaps are

$$nE(d_n) = \frac{n}{p'_{12}} = \frac{n(1 - \delta)}{\lambda(1 - \delta^n)},$$

and

$$nE(g_n) = \frac{n}{p'_{21}} = \frac{n(1 - \delta)}{\mu(1 - \delta^n)}.$$

Observe that since $\lambda, \mu \leqslant 0.5$, $\delta \leqslant 1$. Therefore as n increases, $nE(d_n)$ and $nE(g_n)$ increase also. That is, the inferred lengths of the demes and gaps become greater as the sampling becomes coarser. Consequently the number of distinct demes into which the population is subdivided is automatically underestimated when sampling stations are spaced at a distance more than one unit apart (recall that the unit of distance is the minimum possible length for a gap).

Table 3.1 and Figure 3.11 illustrate the effect. These computations

Table 3.1 The inferred mean run-lengths and mean absolute lengths of demes and gaps when sampling stations are spaced at *n* units. In this example, the Markov matrix of the chain of occupied and empty stations is
$$P = \begin{pmatrix} 0.5 & 0.5 \\ 0.2 & 0.8 \end{pmatrix}$$

Interval between sampling stations n	Mean run-lengths		Mean absolute lengths	
	$E(d_n)$	$E(g_n)$	$nE(d_n)$	$nE(g_n)$
1	2.000	5.000	2.00	5.00
2	1.538	3.846	3.08	7.69
3	1.439	3.597	4.32	10.79
4	1.411	3.529	5.65	14.11
5	1.403	3.509	7.02	17.54
6	1.401	3.503	8.41	21.02
7	1.400	3.501	9.80	24.51
⋮	⋮	⋮	⋮	⋮

should not be taken to imply that it is straightforward to compensate for the errors resulting from coarse sampling. The argument above is based on the assumption that the sequence of demes and gaps along a shore are a realization of a Markov process. There is no reason to expect this to be true in general. The arguments are, however, qualitatively correct whatever the pattern; that is, it is true in general that coarse sampling leads to

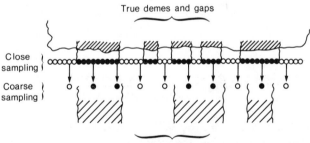

Figure 3.11 Illustration of the effect of coarse sampling on the inferred lengths of the demes and gaps of a littoral species. The upper part of the figure shows the shoreline and the true pattern of demes and gaps as revealed by close sampling at stations one unit of distance apart. The lower part of the figure shows the pattern of demes and gaps inferred from an inspection of every fifth sampling station only.

an overestimate of the sizes of the demes and gaps and an underestimate of their number. In other words, coarse sampling leads one to infer that the pattern on the ground is coarser than it is. The magnitude of the bias is usually impossible to estimate.

Chapter Four

THE QUATERNARY ICE AGE
AND BIOGEOGRAPHY

In Chapter 2 we considered the paleobiogeography of the last 200 m.y. as it has been affected by continental drift. But that chapter did not deal with the recent past, the last 2 m.y., which are special in two ways. First, because of its recency, knowledge about this interval is fairly detailed. Second, during the interval living conditions throughout the world underwent a sudden, dramatic change brought on by the "great ice age," the first ice age since Gondwana was covered by thick ice sheets at the end of the Paleozoic era about 250 m.y. ago. Throughout the enormously long period of time between these ice ages, that is, throughout the Mesozoic and Tertiary, there was no polar ice except occasionally, near the South Pole (see page 44); almost the whole earth had a climate that would be called tropical or subtropical by today's standards and most of the time ice existed only in negligible amounts, in the form of high mountain glaciers.

The period of the "great ice age," the Quaternary, is the period we are now living in. It is divided into two epochs, the Pleistocene and the Holocene; the latter is the present epoch. The Pleistocene-Holocene transition is put at about 10,000 years BP, which coincides roughly with the disappearance of continental ice sheets from most of the earth, though not, of course, from Greenland and Antarctica. The transition from the Pleistocene to the Holocene is not marked by any abrupt climatic change. The current ice age is by no means over, even if it is approaching its end, and there is no reason whatever to suppose that it is.

The dates of the various climatic events in the Pleistocene are very uncertain because different dating methods give discrepant results. It can be said, however, that the Pleistocene epoch is divisible into two parts of which the earlier was preglacial. It was followed by the ice age itself, which has lasted for about the last 0.5 m.y. During this time the ice sheets of North America and Europe have undergone great oscillations. To date, they have waxed and waned at least four times, giving a sequence of glaciations, or glacial stages, each followed by an interglacial stage. The Eurasian ice sheets have always been much more extensive and continuous in the west (Europe) than the east (Asia).

We are now in the fourth (at least) interglacial. During the glacials, the northern hemisphere ice sheets extended south into middle latitudes to cover enormous areas in North America and Europe (see Figure 4.1). In the interglacials they were much smaller, being confined, as at present, to high latitudes. However, these ice caps did not disappear entirely; they have probably existed continuously for over 0.5 m.y.

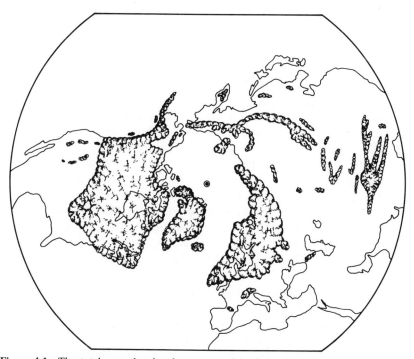

Figure 4.1 The total area that has been covered by ice at one time or another in the Quaternary. Ice cover in the southern Andes, and in the Southern Alps of New Zealand was also heavy. The last glaciation (Wisconsin/Würm) was almost coincident with the glaciated area shown here. (Adapted from Antevs, 1929.)

The glacial and interglacial stages of the two great northern hemisphere continents are known by different names. A table summarizing these stages appears on the endpapers; for Europe are given the older, better-known names that strictly refer to the Alpine glaciations; the high-latitude, northern ice sheets have been given still another set of names which may be found in Kurtén (1971), for example.

The profusion of names for intervals, subintervals, and subsubintervals of time, in areas, subareas, and subsubareas of the world, combine to make Quaternary studies very much a happy hunting ground for specialists. Knowledge concerning the behavior of the ice sheets becomes more and more detailed, of course, as successively more recent events are studied. The glacials themselves are divided into stadials and interstadials, which are lesser oscillations within the greater. For example, the cold of the last European glacial, the Würm, was not constant. Several colder and warmer intervals within it (stadials and interstadials) have been recognized and named. The last three, for instance, were the Lower Dryas stadial (named after the plant genus *Dryas*), which was followed, about 12,000 years BP, by the warmer Alleröd interstadial, which was then followed, about 11,000 years BP, by the Upper Dryas stadial which concluded the Pleistocene epoch.

The Holocene, also, has seen climatic change. Early in the epoch was a warm interval, peaking about 7000 years BP; this warm interval is variously called the Hypsithermal, the Xerothermic, or the climatic optimum. Much more recently, in historic times, has come the so-called "little ice age," the three comparatively cool centuries from about 1550 to 1850 AD.

The Quaternary ice age has had tremendous consequences for the whole biosphere, as may be imagined. The enormous literature on the subject could be classified in a number of different ways, for instance, by region, by time interval, by taxonomic group, or by the nature of the physical changes that have led to biotic change. The present discussion concentrates on the last of these approaches.

The effects on the biosphere of the formation of huge ice sheets can be considered under three heads: effects caused by concomitant worldwide changes in sea level; effects caused by the close proximity of the ice sheet in certain regions; and effects caused by a concomitant climatic change over a wider area. These three topics are discussed in Sections 1, 2, and 3. Section 4 takes a different approach and discusses how all three classes of effect have manifested themselves in combination in one particularly well-studied region, Beringia, the area on either side of, and including, the Bering Strait.

1. THE BIOGEOGRAPHIC EFFECTS OF CHANGING SEA LEVELS IN THE QUATERNARY

According to Fairbridge (1973), the volume of the present-day ice sheets in Antarctica and Greenland is about 24 million km^3. At the height of the last glaciation (the Wisconsin of North America, the Würm of Europe), about 20,000 years BP, the ice volume was 77 million km^3. Thus more than 50 million km^3 of what is now sea water was missing from the oceans at the time of the last glacial maximum, with the result that sea level was 130 m below its position now, in what may be a fairly representative interglacial. In the penultimate glacial (the Illinoian in North America, the Riss in Europe), sea level was probably 160 m lower than at present (Kurtén, 1972). In nonglacial epochs, when the world is practically free of ice, the sea rises to 70 m above its present level (Fairbridge, 1973). Thus the total amplitude of sea level changes between glacial and nonglacial epochs is of the order of 230 m.

These are eustatic changes; that is, they result from changes in the volume of sea water and affect the whole earth equally. In addition, local, isostatic changes occur. These result from localized changes in shape of the ocean basins, caused by crustal warping and other isostatic adjustments in relative height of segments of the earth's crust. There are a variety of geologic causes for crustal warping; two of the most important are the formation and melting of large ice sheets. The weight of an ice sheet depresses the underlying crust, and when the ice melts, the crust rebounds.

The biotic effects of changing sea level are considerable. The most obvious effect, of course, is on the life of tracts of land alternately covered and uncovered by the sea. In addition, the lowering of sea level uncovers migration routes for land organisms and closes migration routes for marine organisms. A rise in sea level, besides having the converse effect on migration routes, brings seashore and shallow-sea habitats into continental interiors. Examples of all these effects are described below. It will be seen that a wide variety of results can stem from an apparently simple, straightforward cause.

Among the important migration routes affected by regressions of the sea are the Bering route between Siberia and Alaska (discussed in Section 4) and the routes linking many of the Indonesian islands to Asia or Australia (see Figure 1.2*). Thus when sea level was low, the Philippines,

*Observe that Pleistocene coastlines would not be expected to follow modern bathymetric contours exactly; the contours undergo repeated slight distortions as a result of crustal warping.

Borneo, Sumatra, Java, and the chain of islands from Java to Timor (the Sunda Islands) became part of mainland Asia; at the same time New Guinea and Australia were linked by a wide land bridge. The two continents themselves were not joined because deep channels separate their respective shelves; Celebes, also, remained isolated (Kurtén, 1972). The zoogeographic boundary well known as Wallace's line (see page 14) corresponds with the permanent strait lying between Asia and Australia. However, because of the glaciations, there was faunal continuity between each continent and many of the islands on its own side of Wallace's line as recently as the late Pleistocene.

The lowering of sea level at the time when the ice sheets were large, exposed strips of what is now submerged continental shelf in many parts of the world; this had the effect of widening preexisting coastal plains. A biogeographic consequence of such a change in Australia has been described by Parsons (1969).

Australian eucalyptus woodlands can be classified as mesic if they have more than 650 mm of rain per annum, xeric if they have less. Woodlands of both kinds form geographically separate blocks, western and eastern, with desert country—the Great Victorian Desert and the Nullarbor Plain—between. The mesic forests of western and eastern Australia have no eucalypt species in common; this suggests that they have been isolated from each other for a long time, probably since the Miocene or before; (the earliest known fossil eucalypt is from the Oligocene). However, drier habitats in the two regions do have species in common; examples are the "mallee" (shrubby) species *Eucalyptus incrassata* and *E. diversifolia* whose ranges are shown in Figure 4.2. This implies that these species had continuous, nondisjunct ranges fairly recently, recently enough for them not to have had time to undergo evolutionary divergence. Parsons (1969)

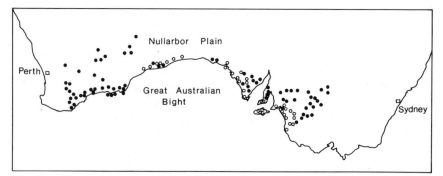

Figure 4.2   The disjunct ranges in Australia of *Eucalyptus incrassata*, ●, and *E. diversifolia* ○. (Adapted from Parsons, 1969.)

hypothesizes that their ranges were continuous along the wide coastal plain, south of the modern coastline, that must have existed during the last glaciation. This coastal lowland, even at its center, would have been less arid than the modern Nullarbor Plain for two reasons: it was farther south, and therefore in a rainier latitude, than the present coast; also, contemporaneous with the existence of ice sheets in the colder parts of the world, there were so-called pluvial periods, with increased rainfall, in the warmer parts. It is therefore probable that conditions suitable for *E. diversifolia* and *E. incrassata* formed a continuous strip south of the modern coastline and that they did not become disjunct until the sea rose and cut through their ranges with the melting of the last ice sheets. The fact that the more mesic eucalypts did not have continuous ranges also may be because the moisture was insufficient or because of an edaphic barrier; (many of the mesic species require soils more acid than those that must have existed on the now submerged coastal lowlands).

In North America, the waning of the last ice sheet (the Wisconsin) caused eustatic and isostatic changes in sea level to take place concurrently (Elson, 1969). Melting of the ice increased the volume of the world ocean and hence caused a eustatic (worldwide) rise in sea level. At the same time, removal of the weight of the overlying ice permitted isostatic uplift of the crust that had previously been downwarped beneath it. Since the disappearance of ice sheets from the North American mainland, the crust has risen by as much as 275 m in some places. However, the isostatic change lagged behind the eustatic change and the net result was the formation of inland seas, when valleys and lowlands in the interior of the continent were inundated. Local geography underwent repeated changes; in the region of the Saint Lawrence River valley and the Gulf of Saint Lawrence, the coastline about 10,000 years ago was as shown in Figure 4.3. The upper Saint Lawrence valley together with the modern Lake Champlain, the Richelieu River valley, and the lower Ottawa River valley were all submerged under the Champlain Sea, which was connected by a wide channel to the Gulf of Saint Lawrence. Thus the Great Lakes and the Atlantic were joined by continuous salt water. Plants previously confined to the Atlantic coastal plain and Atlantic beaches were able to expand their ranges by migrating along these shores and establishing themselves around the shores of some of the Great Lakes, especially around the southern end of Lake Michigan. Crustal upwarping subsequently drained the Champlain Sea, and its shoreline flora disappeared. At present, therefore, a number of species have disjunct ranges; their main ranges are along the Atlantic coast and they have outlying populations around the southern Great Lakes (Peattie, 1922; Cushing, 1965). Examples are *Ammophila breviligulata* (beach grass), *Cakile*

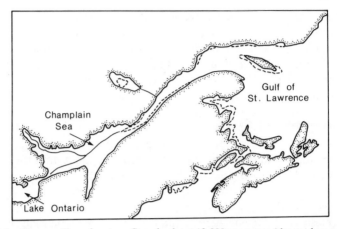

Figure 4.3   The coastline of eastern Canada about 10,000 years ago (the modern coastline is shown by the dashed lines). The sea submerged the valleys of the Saint Lawrence River, the lower Ottawa River, and Lake Champlain, forming the Champlain Sea which linked the Gulf of Saint Lawrence with the Great Lakes. The contemporary location of the ice front is not shown, but see Figure 4.5. (Redrawn from Elson, 1969.)

edentula (sea rocket), and *Euphorbia polygonifolia* (seaside spurge), among beach plants; *Xyris caroliniana* (yellow-eyed grass) of bogs near the sea; and the freshwater aquatic *Utricularia purpurea* (purple bladder-wort). The modern range of *Euphorbia polygonifolia* is shown in Figure 4.4. Besides plants, there are also species of grasshopper with similarly disjunct ranges; two examples are *Neoconocephalus robustus* and *Trimerotropsis maritima,* both Atlantic coastal plain species with inland subspecies.

The changes in sea level caused by the melting of the last ice sheet, and the accompanying changes in sea water temperature, affected the benthic faunas of the continental shelves. An unusual example has been described by Bousfield and Thomas (1975). It concerns the intertidal and estuarine fauna of New England and the Maritime Provinces of Canada (see Figure 4.5). At present there are two sets of species with disjunct ranges. One set, the warm-water disjuncts, are southern animals whose main ranges have Cape Cod as their northern limit, but they are also found in the south-western Gulf of Saint Lawrence around Prince Edward Island; examples are *Triphora perversa* (the three-lined snail) and *Haustorius canadensis* (a sandhopper). The other set of disjuncts, the cold-water disjuncts, are boreal species that are absent from Gulf of Saint Lawrence shores, but reappear farther south between Bar Harbor, Maine, and Saint John, New Brunswick, near the mouth of the Bay of Fundy (which separates penin-

Figure 4.4 The range of *Euphorbia polygonifolia*, seaside spurge. (Adapted from Peattie, 1922.)

sular Nova Scotia from the mainland). Examples are *Buccinum undatum* (a whelk) and *Mysis gaspensis* (a fairy shrimp).

These curious ranges have been ascribed by Bousfield and Thomas to the effect of the Bay of Fundy tides which, at the present day, have amplitudes exceeding those of any other tides in the world; the greatest amplitude is about 17 m. However, between 6000 and 10,000 years ago when sea level was lower, and hence coastal waters shallower, than at present, the amplitude of the tide was comparatively slight. At the same time, the world as a whole was warmer. This was the Hypsithermal period; at its warmest, the time of the "climatic optimum" about 7000 years ago, average sea surface temperatures were 2.5°C higher than at present. Sea water became thermally stratified in summer and though warm-water species could live everywhere in the shallows, cold-water species could survive only where there was sufficient depth of water for

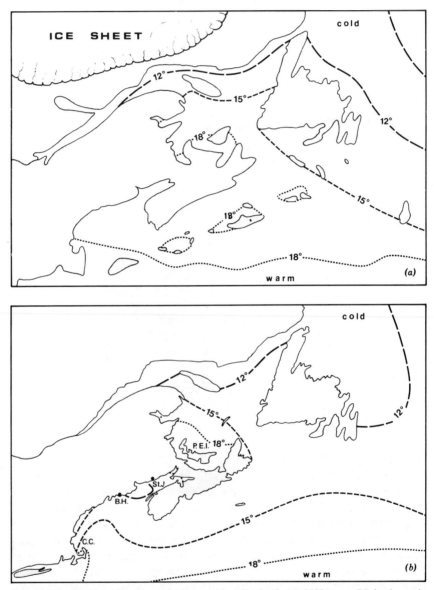

Figure 4.5 Maritime Canada and adjacent New England. (*a*) 9500 years BP in the early Hypsithermal; the ice front was not far north of the Saint Lawrence estuary but the sea in the southern part of the map area was shallow, island dotted, and warm. (*b*) At present. Both maps show summer surface isotherms. (St.J = Saint John; P.E.I. = Prince Edward Island; C.C. = Cape Cod; B.H. = Bar Harbor.) (Redrawn from Bousfield and Thomas, 1975.)

them to migrate below the thermocline in summer. Such a place is the stretch of shore from Bar Harbor to Saint John, where disjunct populations of boreal species still persist.

The disjunct northern occurrences of warm-water species in the Gulf of Saint Lawrence is accounted for as follows. As the ice sheet melted and sea level rose, the shape of the deepening Bay of Fundy basin became progressively modified and is now such as to cause the modern enormous tides. These tides mix shallow and deep water. This prevents thermal stratification of the sea in summer and the formation of warm shallows where warm-water species can survive; consequently, they have been excluded. But they still persist in the Gulf of Saint Lawrence, along the shores of Prince Edward Island and the nearby mainland, because (in contrast to the Bay of Fundy) the Gulf has shallow water and small tides, and thus warm surface water in summer (see Figure 4.5, lower map).

2. NUNATAKS, REFUGIA, AND RELICTS

Wherever the ground was covered by ice during the Pleistocene, all life was destroyed. Every plant and animal now found in the huge areas that were ice-covered (see Figure 4.1) is an invader or the descendant of an invader that arrived in postglacial times. However, within some regions that were almost wholly ice-covered, a few small refugia remained ice-free. They were of two kinds: there were *nunataks,* mountain peaks that protruded above the surface of a surrounding ice field because they were too high to be imbedded; and there were low-level refugia on sea cliffs too steep for ice to rest on them (Porsild, 1969).

In several northern regions there are compact, isolated areas whose modern biotas suggest that they may have been refugia of one of these kinds. Such refugia—if that is what they are—are known in Scandinavia, Iceland, Greenland, and around the Gulf of Saint Lawrence. Concerning many of them, however, there is considerable debate over whether they were in fact refugia.

Two possible refugia whose biotas have been intensively studied are in the mountains of Scandinavia (see Figure 4.6). So far as their floras are concerned, these two areas are very rich in species, many of them rare and several of them endemic. Although the Scandinavian flora as a whole is poor in endemics, such endemics as there are occur mostly in these areas. Plants found in them are described as "unicentric" if they occur in only one of the areas, and as "bicentric" if they occur in both. Among such alpine plant species and subspecies, about 25 are bicentric and 40 unicentric; of the latter 10 are confined to the southern, and 30 to the

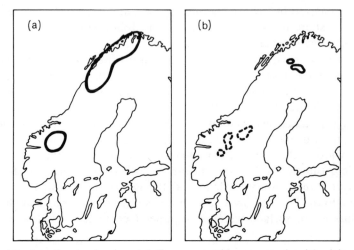

Figure 4.6 Scandinavia showing: (*a*) Two areas with very species-rich alpine floras; species occurring in both areas are "bicentric"; those restricted to only one are "unicentric." (After Gjaerevoll, 1963.) (*b*) The Scandinavian ranges of two species of Lepidoptera. The moth *Acerba alpina* (Arctiidae, tiger moths) occurs in the north (areas with solid outline); the butterfly *Albulina orbitula* (Lycaenidae, "blues" and "coppers") occurs in the south (areas with dashed outline).

northern, of the two centers (Gjaerevoll, 1963). Several "centric" plants (that is, unicentric or bicentric) are widely distributed in northern North America and some beyond, into eastern Siberia, but are present in Europe only in one or both of the Scandinavian "centers." An example is *Rhododendron lapponicum* (Lapland rosebay) whose range is shown in Figure 4.7. As may be seen, its Scandinavian range is bicentric. Two further examples are *Carex scirpoidea* (a sedge) and *Pedicularis flammea* (a lousewort). Both have extensive ranges in North America, and the former in eastern Siberia as well, but in Europe they are found only in the northern of the two Scandinavian centers; that is, they are northern unicentrics. Some insects, also, have unicentric ranges. The Scandinavian ranges of two species of Lepidoptera, one a northern unicentric, the other a southern unicentric, are shown in Figure 4.6*b*.

There is considerable debate as to whether these centric species persisted throughout the last glaciation (the Würm) on nunataks, at the sites where they are now found, or whether they invaded the species-rich centers from offshore refugia on parts of the nearby continental shelf; such refugia would have been dry land when sea level was lower, although now they are submerged.

Gjaerevoll (1963) holds the former view, for two reasons. First, most of

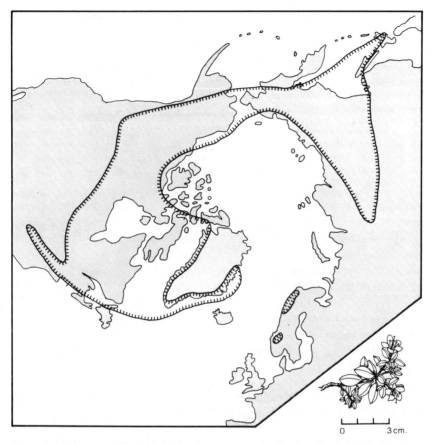

Figure 4.7   The range of *Rhododendron lapponicum* (Lapland rosebay) showing its bicentric distribution in Scandinavia. (Adapted from Gjaerevoll, 1963.)

the centric species have ranges that are entirely inland; if they had invaded the centers from offshore refugia after the ice melted, they would not now be absent from the area between their present ranges and the sea. There are many suitable habitats for them in the coastal mountains. Second, geological evidence shows that the mountain summits in the "centers" were true nunataks; they are covered with weathered detritus ("felsenmeer") that an overlying ice sheet would have been expected to remove. The lower altitudinal limit of the detritus, the so-called "glacial trim line," is probably the line of contact of the ice-field surface and the mountain slopes.

The reasons for rejecting the nunatak hypothesis in favor of the offshore refugium hypothesis are also twofold. In the first place, nunataks, if they existed, must have been extremely inhospitable sites and may have been permanently snow-covered. In any case, conditions on them were undoubtedly as rigorous as those that now obtain on present-day nunataks in Greenland; at most, only a few exceedingly hardy species could have survived on them and such species, because of their toughness, would have been plants that are now wide-ranging. The several alpine rareties now found in the "centers" are unlikely to be survivors from conditions such as these. Secondly, Hoppe (1963) has argued that the geological and geomorphological evidence does not support the nunatak hypothesis. The presence of mountaintop detritus does not necessarily indicate that the mountains were unglaciated; on the contrary, the presence of erratics and glacial striae show that they were. As an explanation for the species-rich floral "centers," Hoppe suggests that they may have been the first areas to become deglaciated and their flora and fauna have thus had a comparatively long time in which to become established.

In North America, in a few places around the Gulf of Saint Lawrence, are found several disjunct populations of species whose main range now is in the far north or in the western Cordillera. It is conceivable that these are relicts that have occupied their present Gulf of Saint Lawrence sites since before the Wisconsin glaciation, on nunataks or low-altitude refugia; even in an area that is known to have been glaciated everywhere (as shown by the presence of glacial till and boulder clay, erratics, glacial striae, and the like), it does not follow that the whole area was under an unbroken ice sheet at any one time. There may always have been some temporarily icefree sites that served as transient stopping places for plants.

Another, more likely possibility (Drury, 1969; Whitehead, 1972) is that these plants immigrated to their present locations from distant sites, mostly in the west, as soon as the ice sheets melted. As melting progressed there must have been, just south of the northward-retreating ice front, a strip of land bare of vegetation and exposed to strong winds. All along this strip, seeds must have been readily dispersed by the wind and been able to establish themselves wherever they chanced to fall because there was no closed cover of competing vegetation. The presence of some of these plants in lowlands that were submerged by the Champlain Sea (see Figure 4.3) shows that these sites, at least, must have been colonized since the ice disappeared; but it does not show whether the colonists came from nearby, possibly shifting refugia where they had endured throughout the glaciation, or whether they were long-distance immigrants from the far west. The latter view is supported by the fact that some western vertebrates, as well, occurred far to the east of their present ranges at various times in the Pleistocene; two examples are *Pedioecetes phasianellus* (sharp-tailed grouse) and *Citellus tridecemlineatus* (thirteen-lined ground squirrel) (Guilday, et al., 1964; Whitehead, 1972). This suggests that dispersal in an east-west direction may have been easy for many organisms before there had been time for climax communities to develop.

Drury (1969) lists 40 plant species and subspecies from the western Cordillera and 5 from more northern communities that are now represented by disjunct populations (of the same or of very closely related taxa) near the Gulf of Saint Lawrence. It is noticeable that nearly all of them are plants of "special" habitats, such as freshwater ponds and lakes, brackish and salt waters, and serpentine and limestone outcrops. Natural selection ensures that such plants are capable of long-distance dispersal; further, the terrestrial ones are usually plants that can succeed in a very wide variety of habitats, and indeed flourish in "ordinary" habitats, if they are protected from competition. Their restriction to unusual, marginal habitats stems from the fact that these are the places where competition is mild because plants unable to tolerate the unusual conditions are absent. A few examples of the species listed by Drury are as follows. Boreal species: *Polygonum acadienses, Ruppia maritima;* Cordilleran (western) species: *Puccinellia macra, Salix brachycarpa, Solidago multiradiata, Vaccinium nubigenum, Antennaria alpina,* and *Honkenya peploides* var. *maxima.*

As the ice front migrated steadily northward at the end of the last glaciation, the vegetation zones to the south of it must have migrated northward in concert (see Chapter 7, page 216). South of the bare zone immediately adjacent to the ice there was a belt of tundra, and south of that a belt of boreal forest. These migrating zones have left some relicts behind.

Thus in what is now boreal forest, there are still a few small regions where trees have yet to establish themselves; shallow, rocky soils have so far kept them out. The vegetation in these small, treeless areas are disjunct southerly outliers of the arctic tundra and contain relict populations of typically northern plants. An example (Porsild, 1958) is *Dryas integrifolia* (mountain avens), a very abundant tundra plant, which has disjunct relict populations at a few places in the Rocky Mountains and also at one isolated site on the north shore of Lake Superior.

Many sphagnum bogs in the boreal forest of North America are relicts of communities whose main range at present is much farther north. Such relict sphagnum bogs are found in undrained kettle-holes, deep hollows in the bogs where detached ice blocks persisted unmelted for many years after the continuous ice sheets had vanished, covered by insulating blankets of sphagnum and sphagnum peat. The ice blocks finally melted to leave kettle-hole lakes, many of which were subsequently filled in by surrounding sphagnum mats. These relict sphagnum bogs provide an environment for other relict populations of associated plant and animal species. For example, Reichle (1966) found that many northern species of pselaphid beetle (Coleoptera; Pselaphidae, mold beetles), whose main range is in the tundra, occur also in sphagnum bogs a long way south of the modern tundra/forest boundary. The pselaphid beetles are weak fliers and require a very moist habitat; this makes it difficult for them to migrate from one bog to another over intervening dry land. Some are now re-stricted to bogs near the southernmost limit of land that was covered by the Wisconsin ice (see Figure 4.8); they were presumably left behind as the ice front retreated farther and farther north and have managed to maintain themselves, as isolated local populations, ever since. Although the environment became warmer and drier in most areas once the ice was gone, cold, damp pockets remained in the vicinity of kettle-holes. Reichle considers that modern conditions in these pockets—low temperatures and extreme wetness—resemble conditions that were widespread adjacent to the ice sheet; in the 35°N to 40°N latitude belt, temperatures at the height of the Wisconsin glaciation (circa 20,000 years BP) were probably 5°C below present levels and precipitation was much greater.

    Not all the Quaternary relict populations surviving as outliers in the

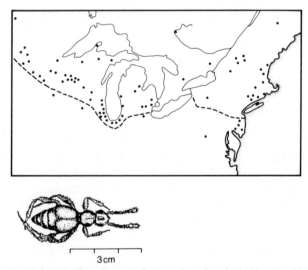

Figure 4.8 The locations of occurrence of 12 species of pselaphid beetles which are found only in sphagnum bogs. The broken line shows the southern limit of Wisconsin ice. (Adapted from Reichle, 1966.) Lower left: *Batrisodes globosus,* one of the species concerned.

"wrong" vegetation zone today are of now northern species left behind in regions south of their present ranges. During the Hypsithermal period the boundaries of the vegetation zones must have been some distance north of their present positions; the zones must have migrated southward as the climate of the northern hemisphere cooled during the past 7000 years, and have no doubt left behind some outlying relict populations in isolated pockets. Some probable examples have been given by Damman (1965) in an account of the elements making up the flora of Newfoundland. For instance, he mentions *Pinus resinosa* (red pine), *Gaultheria procumbens* (checkerberry), and *Chimaphila umbellata* (pipsissewa). These species are found, in Newfoundland, only in restricted, widely separated areas, within which they may be abundant. Damman considers their ranges difficult to account for unless one assumes that the present populations are relicts of populations that were much more widespread in the Hypsithermal period.

3. EFFECTS OF THE GLACIATIONS IN REGIONS FAR FROM THE ICE

Throughout most of the Quaternary, the polar regions have been extremely cold, far colder than they were in earlier, nonglacial periods. There is, therefore, a strong equator-to-pole temperature gradient and it is this that causes the latitudinal zonation of vegetation familiar to all ecologists (see Figure 7.2). The vegetation that now covers once-glaciated areas has been there only since the ice melted, of course, and two factors that could cause zonation are conceivable. The first is the gradient in environmental conditions already mentioned; this is almost certainly the chief cause. The second, possibly contributory, cause is that it has taken time for plants to disperse into, and colonize, newly icefree areas; plants with different speeds of dispersal have reached "fronts" at different latitudes and the resultant sorting effect may have reinforced the environmentally caused zonation.

These two causes of zonation are difficult to disentangle, however. Temperatures have not risen consistently in the last 10,000 years; there have been fluctuations at several frequencies. It is conceivable that every plant species reacts individually to these fluctuations, with the northern limit of its range shifting alternately north and south at a rate specific to itself, as temperatures rise and fall. Advances and retreats would lag by different amounts behind the environmental changes that caused them. And, for all but pioneer species capable of colonizing bare mineral soil, successful northward advance of a species would depend not only on climate but also on the vegetation which had gone before it and lay ahead of it and which would have created suitable conditions for it. Therefore the zonation we observe is controlled by the fluctuating rate of temporal change of the latitudinal temperature gradient, by the rates of dispersal of different species, and by the rate of ecological succession. Succession itself proceeds at a rate controlled by temperature. The effects of these interacting feedbacks may well be inextricable, but attempts to model them might lead to recognition of the relative importance of the several causes of zonation.

Interesting, and perhaps explicable, variations in the widths of the vegetation zones exist. Thus all zones are displaced southward in the neighborhood of cold seas. The effect is evident in the Labrador–Newfoundland region and in the neighborhood of the White Sea, and is especially clear around Hudson Bay (Figure 4.9). Savile (1968) has discussed the effect of Hudson Bay on the vegetation of the lands around it. In this region, tundra vegetation extends into the lowest latitudes at which it is found at the present day.

As is well known, a large body of water acts as a heat reservoir making the climate of its shores less extreme, and hence milder in winter, than the climate of inland continental regions at the same latitude. This moderating of seasonal temperature variation will not permit nonhardy plants to grow, however, unless the average around which the moderate variations take place is high enough. In the neighborhood of Hudson Bay, it is not. The ice pack in the bay persists until midsummer and even after it melts the water remains very cold. Summer warmth thus comes far later in the

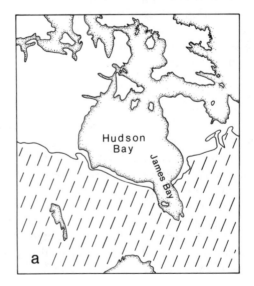

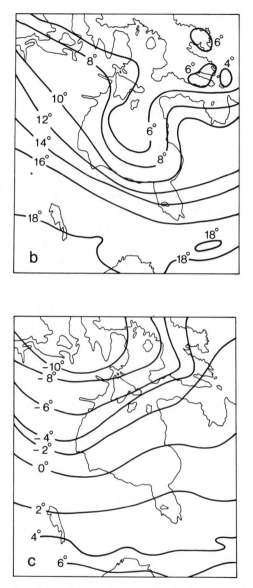

Figure 4.9 Hudson Bay, James Bay, and surrounding land. (*a*) Boundary between boreal forest (hatched) and tundra. (*b*) Mean daily temperature in July (degrees Celsius). (*c*) Mean daily temperature in October (degrees Celsius).

year to the shores of the bay than to inland areas; the air is still very cold in the season when daylight lasts long enough for most plants to make their best growth. And although the heat-retaining property of the bay has the effect of delaying fall freeze-up somewhat, this delayed "warm" (above freezing) period comes too late to be useful for plant growth since by the time it arrives the number of hours of daylight per day is rapidly diminishing (Savile, 1968).

Although the northern limit of trees (equivalently, the southern limit of tundra) bends far to the south because of the influence of Hudson Bay, it does not lie south of the extreme southernmost part, James Bay. The tundra zone is thus divided into an eastern and a western section. Presumably when polar ice was more extensive than it is now, the tundra zone must, for a period, have been continuous around the head of the nested bays. The modern distributions of mammal species of the tundra, however, suggest that the tundra corridor linking the eastern and western sections must have been short-lived; or, what comes to the same thing, that the boreal forest must have advanced northward very rapidly after the ice disappeared.

The reason for supposing this to be true was given by Macpherson (1968). The mammals that colonized the tundra after the melting of the ice came chiefly from Beringia (the Bering Strait region plus Alaska and Yukon); northern Greenland was a comparatively unimportant secondary source. At present there is a pronounced contrast between the mammal faunas of the two sections of tundra; although nine species occur on both sides of Hudson Bay (examples are *Rangifer arcticus*, barren ground caribou; *Alopex lagopus*, arctic fox; *Lepus othus*, tundra hare), there are seven species that are found only to the west of the bay (examples: *Citellus parryi*, arctic ground squirrel; *Clethrionomys dawsoni*, tundra redback vole), and one species (*Dicrostonyx hudsonius*, Hudson Bay collared lemming) only to the east of the bay. The bay itself forms a barrier to east-west dispersal of terrestrial animals and they are now prevented from migrating around the head of the bay by the boreal forest zone which constitutes an environmental barrier. Macpherson (1968) considers that the mammal faunas of the eastern and western sections of the tundra zone would resemble each other more closely than in fact they do if they had been linked by a tundra corridor for a longer period.

The preceding paragraphs serve to emphasize how numerous are the complicating factors that crop up when one considers how the modern geographic ranges of plant and animal species have been affected by the Pleistocene glaciations. They also emphasize that throughout the past 0.5 m.y. climate, physiography, and vegetation have all undergone continuous change at rates far more rapid than those that prevailed in the previous 200 million or more nonglacial years. Many of the complexities of modern biogeography stem from this fact.

It should now be noticed that those parts of the world that were ice-covered at one time or another in the Pleistocene are not the only areas that have been affected by the ice. While the great ice sheets expanded and shrank, climates elsewhere varied enormously. The biotic history of the unglaciated areas was governed by these variations. The presence of huge ice sheets in high latitudes caused worldwide changes in atmospheric circulation (Flint, 1957). In each hemisphere, the belt of eastward-moving cyclonic storms was shifted toward the equator when the ice sheets expanded and then shifted back into high latitudes, as today, when they shrank to comparatively small size. Thus at times of maximum glaciation, many areas in the modern desert belt received ample precipitation; they experienced so-called "pluvial" periods. At the same time, equatorial regions where rainfall is now heavy experienced dry periods.

These effects were complicated by the presence of mountain ranges, by transgressions and regressions of the sea as the volume of the oceans alternately increased and decreased, and by the formation and drying up of "pluvial lakes"; these lakes filled and evaporated in response to melting and freezing of the ice sheets, to changes in precipitation, and to changes in the evaporation rate which itself rose and fell in response to temperature and degree of cloudiness (Flint, 1957). At times when the ice sheets were extensive, the climatic zones between the polar frigid zones were telescoped together. Climatic gradients were steep. The details are extraordinarily complicated and because local effects were everywhere important, probably no general rules apply. In South America, for example, according to Vuilleumier (1971), at any time in the Pleistocene the climatic processes on one side of the Andes were the reverse of those on the other side.

The biogeographic effects of these frequent and pronounced climatic changes have been well studied in South America. The chief effect has been the alternate fragmentation and reunification of tracts of habitat. Thus in wet periods, the Amazon basin was entirely filled with continuous, dense, rain forest, and in dry periods these forests contracted until they were reduced to a few small isolated "islands" in a "sea" of more xeric vegetation (Figure 4.10) (Haffer, 1969).

Similarly, the altitudinal vegetation belts on the slopes of the Andes moved up and down as the climate warmed and cooled. The humid grasslands ("paramo") lying between timberline and the permanent snowline in the northern Andes behaved in this manner. In warm periods, the paramo is restricted to narrow belts high in the mountains, separated

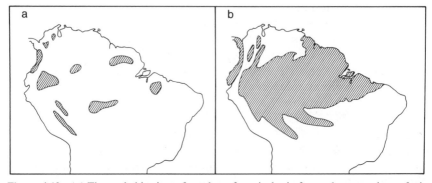

Figure 4.10 (*a*) The probable sites of patches of tropical rain forest that served as refugia for forest birds in dry periods in the Pleistocene. (After Haffer, 1969.) (*b*) The current extent of tropical rain forest in northern South America.

from one another by intervening montane forest. In cool periods, the paramo is forced down to lower altitudes, causing its separate parts to coalesce (Vuilleumier, 1971).

Both the Amazonian rain forest and the paramo have therefore been fragmented repeatedly. The forest shrank to isolated fragments in dry periods and reexpanded in humid periods. The paramo shrank to fragments in warm periods and reexpanded in cool periods. The effects of these changes on the speciation patterns of many taxa of birds, plants, and lizards have been well studied (Haffer, 1969; Vuilleumier, 1971). Divergent evolution in the isolated islands of a fragmented habitat permits new species to appear (see Chapter 3, page 91). Then, when the fragments coalesce, the separated populations come into contact again; although some of the newly evolved species may go extinct because of competitive exclusion, others, if they have become adapted to different microhabitats, can coexist successfully. In plants, moreover, secondary contact often favors the appearance of still more new species as a result of alloploidy.

Thus it is very likely that the fauna and flora of tropical South America owe their well-known richness to climatic fluctuations during the Quaternary, that is, during the last 1 or 2 m.y. These fluctuations probably exceeded, in both rapidity and amplitude, any that took place in the preceding 200 m.y.

Notice, however, that if alternating fragmentation and coalescence of populations is to result in an "explosion" of new species, it is necessary that the frequency of the fluctuations be matched to the evolutionary rates of the organisms. The optimum frequency, from the evolutionary point of view, is obviously that which allows diverging populations just enough time to become reproductively isolated from one another before secondary contact ensues. Therefore, it is important to consider evolutionary rates in different groups of organisms and whether, as climate fluctuated, well-differentiated species had time to evolve. Haffer (1969) has discussed this topic. He speculates that 20,000 to 30,000 years probably suffices for the evolution of a new species of small passerine bird, and that a period of the order of a few hundred thousand years may be necessary for larger birds. If this is so, it follows that there has been time for repeated speciations among birds since the end of the Tertiary. Probably the same conclusion holds for the insects, amphibians, reptiles, and mammals of the Amazonian fauna.

Elsewhere in the world, climatic change in the Quaternary has not always led to diversification. On the contrary, there has been a catastrophic wave of extinctions in the large vertebrates, especially the large mammals. Such famous animals as the woolly rhinoceros, mastodont, mammoth, giant beaver, cave bear, giant ground sloth, giant deer, American camel, and saber tooth cat have disappeared. Extinctions have continued into historic times with the loss of the aurochs and, among birds, of the moa and the dodo (Kurtén, 1972). The cause of these extinctions is hotly debated (Martin and Wright, 1972). In particular, there is disagreement over the degree to which human activity has been responsible. Many of the extinctions have occurred during the period in which *Homo sapiens* appeared and multiplied. But climates, and hence habitats, have been fluctuating markedly at the same time. It is difficult to judge how the blame should be apportioned between these concurrent causes. Some consider that what Martin (1967) has called "prehistoric overkill" was the dominant factor; others consider climatic change to have been more important. Kowalski (1967) argues that man was largely responsible, although less through hunting than through destruction of habitat by deforestation and the like. Even so, in southeastern Australia, a great many now extinct giant mammals coexisted with man for at least 7000 years and were still extant as recently as 26,000 years ago (Gillespie et al., 1978).

Climatic fluctuations and the concomitant expansions and contractions of different kinds of habitats must have affected the large mammals profoundly. Although, as discussed in earlier paragraphs, these causes can bring about speciation "explosions" in some taxonomic groups, in other groups the effect may be exactly the opposite. Thus in a dry period, surviving patches of rain forest may be too small to serve as refugia for large mammal species, whose isolated, relict populations may dwindle to the point of extinction. At the same time, expansion of the savannah habitat at the expense of rain forest enlarges the available area for grazing ungulates and their predators, and also permits them to migrate long distances which they could not do if barriers of dense forest blocked the way. Thus expansion of habitat allows populations to build up, while at the same time increased migration and the consequent mingling of species-populations lead to competitive exclusion (see page 56).

This discussion should make clear that a single cause can lead to diametrically opposite outcomes in different kinds of organisms. Only with the aid of hindsight can we now interpret what probably happened. Indeed, because of the large number of interacting population processes that respond in various ways to environmental changes, it is fairly easy to invent a plausible chain of events leading from almost any starting point to almost any end point. This makes qualitative reasoning unsatisfactory. Quantitative studies are needed if we are to judge between many attractive, but conflicting, theories about the responses of different segments (taxonomic and geographic) of the biosphere to the Quaternary ice age. To understand what is going on (I use the present tense deliberately) it is necessary to assess correctly the relative magnitudes of all the many forces that have operated and continue to operate.

4. THE BIOGEOGRAPHIC HISTORY OF BERINGIA

In this section we consider the biogeographic history of one particular region, Beringia (Bering Strait with adjacent eastern Siberia and Alaska) where intensive studies have been made. It is desirable to begin the story early in the Tertiary period, when the climate was mild, and trace the effects of the climatic deterioration that set in probably no later than the early Pliocene. The glaciations of the Pleistocene are the culmination of a climatic trend that had been in progress for several million years and there was no discontinuity in the trend at the time of the Pliocene-Pleistocene transition (equivalently, of the Tertiary-Quaternary transition).

Through the period we shall be considering, the level of the sea rose and fell from both eustatic and isostatic causes. As a consequence, the present Bering Strait for part of the time has been a channel of the sea (as it is today) separating eastern Siberia from Alaska and joining the Arctic and Pacific oceans; for the rest of the time it has been dry land, forming a migration route between Siberia and North America for terrestrial organisms but acting as a barrier to the movements of marine organisms between the oceans.

While these changes in sea level were going on, the climate and hence the vegetation were changing concurrently. There was worldwide climatic change; and at the same time, as the sea transgressed and regressed, the climates in areas that were dry land (for all or part of the time) varied from maritime to continental. The vegetation has varied correspondingly. At different times, there were deciduous forests, coniferous forests, xeric grasslands (steppe or prairie), tundra, and local ice sheets. Beringia was never covered by continuous ice (see Figures 4.1 and 4.12).

The animals, resident and migrant, have been affected by both climate and vegetation. Thus migrating herds of grazing ungulates must have grass on which to graze, and adequate grassland develops only where the climate is simultaneously warm enough for the grasses to grow and dry enough to prevent forest becoming established.

In the Quaternary, at the height of the glaciations, there were three contrasted "terrains" in the area: sea, dry land, and ice. Icefree land and ice sheets are obviously very different as dispersal routes for terrestrial animals since ice is an impassable barrier to all but fairly large mammals and (possibly) some insects and other small arthropods.

The history of Beringia thus provides good examples of a wide range of biogeographic phenomena. The area is important both in its own right and as a link between the Palearctic and Nearctic terrestrial biogeographic regions (see Figure 1.1). It is also a link between the Pacific and Arctic Oceans. Raven and Axelrod (1974) consider that it has been more important than the North Atlantic as a land route for migrating plants since about the middle of the Eocene (see page 31). There have probably been no land connections around the North Atlantic since the early or middle Tertiary but Beringia has probably been a land bridge since very early in the Tertiary (Hopkins, 1967).

Let us now consider events from that point on, drawing upon the accounts of Hopkins (1967), Wolfe and Leopold (1967), Colinvaux (1967), and Flerow (1967).

Beringia in the Tertiary

Until the late Miocene, the climate of Beringia was mild and there was a wide land bridge between Asia and North America. The land bridge was forested; its forests were part of an unbroken belt of mesophytic deciduous forest stretching from eastern Asia, including Japan, around the northern shores of the Pacific Ocean and then south and east as far as southeastern North America. Representative tree genera were *Carya* (hickory), *Juglans* (walnut), *Corylus* (hazel), *Fagus* (beech), *Quercus* (oak), *Ulmus* (elm), *Liquidambar* (sweet gum), *Platanus* (plane or sycamore), *Acer* (maple), *Ilex* (holly), *Tilia* (lime or basswood), and *Nyssa* (tupelo). Most of these genera still have representatives in the warmer parts of their earlier ranges but have been absent from Beringia ever since the climate became too cold for them in the late Tertiary.

These currently disjunct tree ranges are matched by those of the phytophagous insects that feed upon them. For example, there is a strong similarity between eastern Asia and eastern North America in their representatives of the Cerambycidae (Longhorn beetles), a family of fairly host-specific wood borers that attack trees (Linsley, 1963).

There are also a number of well-known shrubs and herbs with disjunct ranges, each having one part of its range in eastern Asia and the other part in eastern North America. These are plants that still thrive in the deciduous forests that persist at each end of the extensive deciduous forests of the Tertiary, the so-called Arcto-Tertiary forests. The disjunct populations are usually congeneric rather than conspecific (Li, 1952). For example, the genera *Epigaea* (trailing arbutus) and *Stewartia* (a genus of shrubs of the tea family) (see Figure 4.11) have disjunct ranges of this kind. (Further examples are given in Chapter 9; see page 270.)

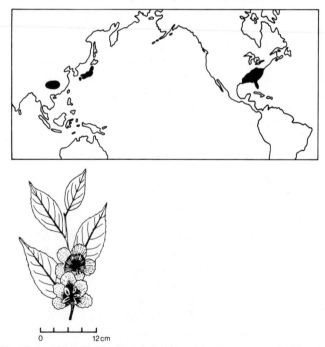

Figure 4.11 The modern distribution of the genus *Stewartia* L. (family Theaceae, the tea family). (Redrawn from Hutchinson, 1926, where the name is spelled *Stuartia*.)

In considering the taxonomic rank of taxa with disjunct ranges, it is important to bear in mind that, in the opinion of some taxonomists, botanists are more given to taxonomic splitting than are zoologists. Often a group that a zoologist would call a subspecies is treated by a botanist as of specific rank (Stegmann, 1963; and see page 42); (opinion is divided on whether or not this is so). Hence the fact that among species common to North America and Siberia there are very few coniferous tree species but many forest bird species may not conflict with the fact that evolution is very much slower in gymnosperms than in passerine birds (pages 41 and 129). According to Stegmann, the paradox is only apparent, and stems from the contrasting taxonomic procedures of botanists and ornithologists. Of course, it could be argued that, since birds can fly, the fact that many species have trans-Beringian ranges is unsurprising. But it is probably a mistake to assume that land birds extend their ranges by crossing arms of the sea, even narrow ones. Nonmigrant land birds never fly over the sea voluntarily and it is unlikely that occasional storm-blown strays could found colonizing populations (Stegmann, 1963; Diamond and May, 1976).

The climatic cooling that created a gap in the great arc of deciduous forest around the North Pacific came at the end of the Miocene. In Beringia, deciduous forest was replaced by boreal coniferous forest, in which the following genera occurred: *Picea* (spruce), *Abies* (fir), *Pinus* (pine), *Tsuga* (hemlock), and *Larix* (larch). In addition to the conifers, as in modern far-northern latitudes, were *Betula* (birch), *Populus* (poplar), *Alnus* (alder), and *Salix* (willow).

At the same time as this vegetation change was taking place, tectonic movements caused the appearance of a trans-Beringian seaway (see Figure 4.12). It formed a barrier to terrestrial organisms but a corridor for marine ones, and there was a faunal exchange between the Atlantic and Pacific via the Arctic Ocean. Animals known to have migrated include mollusks and Pinnipedia; before the opening of the seaway, sea lions and walruses were restricted to the Pacific side, and true seals (family Phocidae) to the Atlantic side, of the land bridge (Hopkins, 1967). The number of mollusk species moving from the Pacific into the Atlantic greatly exceeded the number going the other way. Two probable reasons for this are: first, there was a far bigger pool of species on the Pacific side; second, there is an eastward circumpolar current in the Arctic Ocean, so that dispersal tends to be counterclockwise round the pole, in a west-to-east direction. Migrants carried by the current have a far longer journey, through colder seas, if they are on the Atlantic-to-Pacific route, along the Arctic coast of Siberia, than if they are on the Pacific-to-Atlantic route,

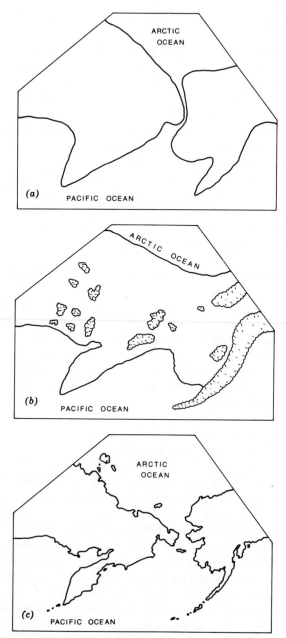

Figure 4.12 Beringia. (*a*, *b*) Speculative reconstructions of Beringia at the end of the Miocene and at the maximum of the Wisconsin glaciation. (Adapted from Hopkins, 1967.) (*c*) The modern coastline.

through the Canadian Arctic archipelago. Thus there was a far greater number of successful migrants by the latter route.

Fossil evidence suggests that the Bering land bridge was in existence again in the middle and late Pliocene and provided an overland migration route for such forest animals as beaver and flying squirrel. Besides coniferous forests, which now had different species-compositions on the two sides of the "strait" (a land bridge at the time), the land was probably covered with muskeg (sphagnum bogs) alternating with woodlands of willow, alder, poplar, and birch.

Beringia in the Quaternary

Since the beginning of the Quaternary, conditions have been changing in Beringia—as they have everywhere else in the world—much more rapidly than before. The alternately falling and rising sea level uncovered and resubmerged the land bridge on at least three occasions (Flerow, 1967) and perhaps as many as six (Hopkins, 1967).

Colinvaux (1967) has reconstructed the vegetational history of Beringia in the Quaternary by means of pollen analysis. The area was never completely ice-covered (see Figures 4.1 and 4.12). As the climate cooled, during the glacials, and warmed, during the interglacials, the northern range limits of most plant species shifted alternately south and north. It is likely that forests finally disappeared from the area some time in the Yarmouthian interglacial (the interglacial preceding the penultimate, Illinoian, glaciation), when the vegetation seems to have been much as it is at present.

During the last two glaciations, the Illinoian and the Wisconsin, the climate was evidently cooler than it is now, with cooler, shorter summers. This is shown by the successive positions, as judged by fossil pollen, of the northern limit of dwarf birch (*Betula glandulosa*), a locally abundant dwarf shrub of the low arctic.

During the glaciations, that is, when the ice sheets were large, the sea level was low, the land bridge was uncovered, and the climate was more continental than it is at present now that the strait forms a wide seaway across Beringia. It was therefore more arid, and this is shown by the vegetation of unglaciated parts where the tundra contained a large admixture of xeric grasses and xerophytes such as *Ephedra, Artemisia* (sage brush), and various members of the Chenopodiaceae (the goosefoot family).

These grasslands provided the grazing required by the large herds of ungulates that migrated across the land bridge during the glacials. A very large number of mammal species entered North America from Siberia during the Pleistocene and have contributed to the present close resemblance between the Palearctic and Nearctic mammal faunas (see page 10). The number of migrants going the other way, from North America to Asia, was much smaller, probably because there was a much less extensive ice cover in Asia (see Figure 4.1), and hence a larger pool of mammal species. Besides grazers, there were browsers and predatory carnivores. The mammals concerned required a wide variety of habitats and these Beringia must, at one time or another, have provided.

The migrating mammals fall into two classes (Flerow, 1967): (1) those that are still present in Alaska and northern Canada, and are similar or identical to Siberian species; (2) those that dispersed southward into temperate North America and in doing so diverged evolutionarily from their ancestors and became adapted to new environments. Extant species in these groups include the following (the nomenclature follows Flerow, 1967).

In group 1:

Ovis nivicola, snow sheep
Ovibos moschatus, musk-ox
Alces alces americanus, moose
Cervus elephas canadensis, wapiti
Ursus arctos middendorffi, brown bear

Gulo gulo luscus, wolverine
Canis lupus, wolf
Vulpes vulpes, red fox
Lynx lynx, lynx
Lepus timidus, arctic hare

In group 2:

Bison bison, plains bison
Ovis canadensis, bighorn sheep
Oreamnos montanus, Rocky Mountain goat
Ursus horribilis, grizzly bear

Ursus americanus, black bear
Canis latrans, coyote
Vulpes macrotis, kit fox
Lynx rufus, bobcat

Many species that are now extinct in North America also traversed Beringia in the Pleistocene, or were descended from ancestors that had made the crossing. For example, *Camelops* (camel), *Equus* (horse), and *Lynx issiodorensis* (a puma-like lynx) migrated westward, from North America into Asia. *Mammuthus* (mammoth), *Mylohyus* (long-nosed peccary), *Homotherium* (scimitar cat), *Smilodon* (saber tooth cat), *Euceratherium* (shrub ox), *Mammut* (mastodont), several species of *Bison*, and a great throng of others migrated eastward, from Asia into North America (Repenning, 1967). These were the victims of the mass extinctions of the Pleistocene (see page 129) which in North America, according to Kurtén (1972), in "perhaps no more than a millennium or two, swept away a host of animal species."

One of the last of the mammals to make the crossing from Asia to North America was *Homo sapiens*. It is thought (Jelinek, 1967) that the migration may have been as recent as 17,000 to 19,000 years BP, soon after the maximum of the Wisconsin glaciation. The migrants were probably of Upper Paleolithic culture and were, no doubt, experienced hunters. The only available food was the abundant game present in the steppe and tundra through which they traveled. Jelinek considers that this hunting was the chief cause of the great wave of extinctions among the other recently arrived New World mammals. He argues that the herbivorous mammals were not adapted to withstand human hunting pressure. The populations that had migrated to North America were removed from human contact and evolved accordingly: wariness in the presence of *H. sapiens* was not selectively favored in North America as it had been in Asia. Thus the "game" animals, even though they were skilled in avoiding carnivore predators, were very vulnerable to human predation. As mentioned earlier (page 129), there is no unanimity of opinion on the cause or causes of the Pleistocene extinctions. Men living now can only regret that the magnificent ice age megafauna, through whatever cause, has been all but wiped out.

Chapter Five

THE BIOGEOGRAPHY OF MARINE ORGANISMS

Thus far in this book, terrestrial organisms have been used in examples of most of the biogeographic phenomena described. We now turn to a consideration of marine organisms. The marine biosphere, no less than the terrestrial biosphere, has been affected by the rearrangement of the continents resulting from movement of the earth's crustal plates, and by the continual variation of the world's climate, especially during the current ice age. These causes have influenced the course of evolution, and the patterns of dispersal and migration, of all marine organisms. The organisms themselves range from diatoms to whales, from littoral algae exposed to the air for several hours in each tide cycle to the deep-sea benthos in perpetual darkness at depths of 10,000 m and more.

There are three chief contrasts between the marine and terrestrial environments, which affect their biogeography profoundly. First, the world ocean is continuous and composed of a fluid medium in constant movement; different parts of it can therefore never be as isolated from each other as can the separate parts (continents and islands) of the world's land surface. Second, the "vegetation" responsible for primary production in the open sea is phytoplankton, consisting of minute floating organisms (diatoms, dinoflagellates, and the like) dispersed through the water; therefore it does not, as does the vegetation of the land, form a structured environment. Third, the ocean is a three-dimensional realm, whereas the land is effectively a two-dimensional one; its local properties, as they affect the organisms present, vary with depth as well as with geographic location.

The contrasts between marine and terrestrial biotas are equally striking. Taxonomically, marine animals present far more variety than terrestrial animals; in the fauna of the sea, all animal phyla are represented.

There are also very pronounced contrasts between marine and terrestrial communities as components of ecosystems. Primary productivity and plant biomass per unit area are both far less in the sea than on the land. Thus (Ehrlich, Ehrlich, and Holdren, 1977) the mean net primary productivity, in grams of carbon fixed per square meter per year, is 144 for marine, and 324 for terrestrial systems. Primary productivity per unit area in tropical rain forest is 15 times as great as in the open ocean (McLean, 1978). The areal density of plant biomass responsible for the photosynthetic fixation of carbon, in grams of carbon per square meter, is 1630 for marine, and 5550 for terrestrial systems. The total quantities of plant matter in the marine and terrestrial biospheres—circa 830×10^{15} g in each—turn out to be closely similar, taking account of the fact that the proportions of the earth's surface covered by ocean and land are 71% and 29% respectively. The ratio of plant to animal biomass, or equivalently of autotrophic to heterotrophic material, is far less in marine than in terrestrial systems; reflecting this difference is the fact that marine systems have a much greater number of trophic levels.

These ecological contrasts between the marine and terrestrial biospheres should be borne in mind as we consider the biogeography of the sea. In Chapter 1 (page 5) it was remarked that both biospheres could be classified in two ways, biogeographically (that is, taxonomically, on the basis of floras and faunas) and ecologically, but that marine biogeographic regions tended to be less well defined than their terrestrial counterparts. Methods of classifying the marine biosphere are discussed in Section 1 of the present chapter.

Section 2 discusses the history of the marine biosphere through geological time, as it affects present-day biogeographic distributions. Section 3 discusses the latitudinal zonation of the marine biosphere, which is better defined and more obvious than that of the terrestrial biosphere. Section 4 discusses dispersal in the sea. Evolution and speciation in the sea are discussed in several contexts, in different sections. As will be seen, one could thread one's way from one topic to the next by a multitude of different but equally acceptable routes; almost any permutation of topics would have something to recommend it. This is because in marine biogeography, as in terrestrial, every phenomenon is influenced by a vast array of causes, some worldwide and long-lasting, others local and ephemeral, some affecting all organisms indiscriminately, others affecting only a specific few.

1. CLASSIFICATION OF THE MARINE BIOSPHERE

The oceans of the world are made up of two realms (see Chapter 1, page 14), the continental shelves and the deep ocean. Treated as biotic environments, these realms can alternatively be labeled the *neritic zone* and the *open sea*.

Both realms provide two radically different habitats, the water itself and the sea floor beneath. These are occupied by *pelagic* and *benthic* organisms respectively.

Pelagic organisms belong to three categories: the *nekton,* or active swimmers; the *plankton,* or passively drifting organisms; and the *neuston,* those organisms that permanently inhabit a thin layer of water at the surface.

Benthic organisms are of two kinds, those that live on the sea floor (epifauna), and those that live within it (infauna), buried in the sediment.

The classification of the neritic zone is comparatively straightforward. One possibility (Sylvester-Bradley, 1971) is to treat the continental shelves as extensions of the land and class them as parts of the terrestrial biogeographic regions on whose borders they lie.

A finer classification has been proposed by Briggs (1974) who recognizes no fewer than 23 regions in five sets. These sets of regions are: the Arctic and northern cold temperate with five regions; the northern warm temperate with four regions; the tropical with four regions; the southern warm temperate with five regions; and the southern cold temperate and Antarctic with five regions.

Before considering how the two marine realms can best be classified, it should be recalled (page 15) that, in general, marine taxa have much bigger ranges than terrestrial taxa; consequently, regional faunas of the sea are less distinct from each other than regional faunas of the land. However, neritic faunas are better characterized taxonomically than are faunas of the open sea; thus whereas neritic faunas often have distinctive families and genera of certain orders of animals, pelagic faunas of the open sea may differ from each other only at the species level (Ekman, 1953; McGowan, 1971).

Pelagic Faunas of the Open Sea

There is considerable doubt as to whether the waters of the open sea can and should be biogeographically classified. Physical oceanographers recognize a number of distinct water masses (their distinguishing properties are discussed below) but there is disagreement on whether each water mass has a distinctive pelagic fauna, which might be called its own faunal identity (Briggs, 1974). According to Funnell (1971), the "water masses tend to be characterized by their own biota" and much evidence has been gathered that suggests this is true of many plankton species. McGowan (1971) considers that the ranges of the majority of pelagic species coincide with water masses and that plankton organisms tend not to stray from one water mass to another. Although there is a certain amount of "leakage," each population is, for the most part, confined to its own water mass.

However, many organisms appear to have distributions that are much less restricted (Briggs, 1974). For example, of 35 species of pelagic chaetognath (arrow worms) occurring in the open sea, 32 are found in several water masses, in more than one ocean. The same is true of 30 out of 33 species of pelagic polychaetes (bristle worms) and 35 out of 48 species of euphausids (krill). Species of pelagic fish living at considerable depth often have ranges whose northern and southern boundaries coincide with the northern and southern boundaries of water masses, especially if the latter are marked by sharp temperature discontinuities; but the ranges of these fish species may extend far beyond the limits of any one water mass in the east-west direction (Briggs, 1974).

Physically, each water mass is a body of water that retains its identity and behaves as a coherent entity so far as circulation is concerned. It constitutes a "gyre," a body of water that circulates as a unit (Pickard, 1975).

Every element of volume of water in the sea has its own temperature and salinity. If water temperature and salinity are measured at a sequence of depths at one site, the measurements provide data for a temperature-salinity graph, or *TS diagram* (see Figure 5.1). It is found that particular TS relationships are typical of particular regions of the ocean. Thus all TS diagrams from one such region, say the eastern North Pacific, resemble each other fairly closely and may be represented by an average or typical curve; this curve clearly differs from the typical curve for another region, say the eastern South Pacific (Figure 5.2). The oceans can therefore be divided into regions each of which shares a common TS relationship. These regions have a north-south extent of between 30° and 50° of latitude. Their east-west extent may embrace a whole ocean, as do the so-called "North Atlantic Central Water" and the "Indian Equatorial Water," or half an ocean (for example, the "Western North Pacific Central Water"), or may be circumglobal (for example "Subantarctic Water" and, to the south of it, "Circumpolar Water"). These and other "waters" are all water masses encountered in the uppermost layer of the ocean (excluding the very uppermost 200 m which are treated as lying above the water masses; see below).

Water masses are not defined only in terms of their areas, however. The

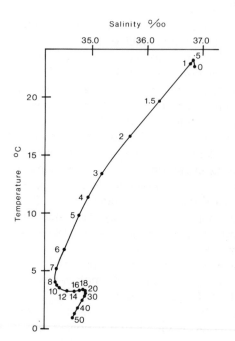

Figure 5.1 A TS diagram for 22°S, 24°W, a point in the tropical South Atlantic. The points on the curve show depths, in meters × 100. Observe that water properties change very much more slowly with depth as depth increases; this is shown by the fact that the points on the curve are closer at greath depths even though the depth interval between points becomes larger. (Adapted from Neumann and Pierson, 1966.)

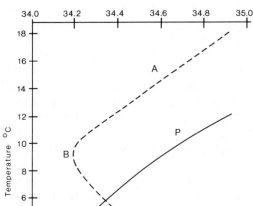

Figure 5.2 Two typical TS diagrams. *Solid line,* Eastern North Pacific Ocean. *Dashed line,* Eastern South Pacific Ocean. Depths are not shown. The capital letters label the layered water masses. From the surface down these are: in the north: A, Eastern North Pacific Central Water; B, Eastern North Pacific Intermediate Water; C, Pacific Subarctic Water. In the south: P, Eastern South Pacific Central Water; Q, Antarctic Intermediate Water; R, Circumpolar Water; S, Antarctic Bottom Water. (Adapted from Williams, 1962.)

ocean is also layered, and a given water column may sample two, three, or four separate water masses all forming layers one on top of another. The shape of a TS diagram gives evidence of the existence of these layers, and also of the degree to which they are mixed. Any one water mass has a fairly restricted range of temperatures and salinities and is represented by a short portion of a TS curve; moreover temperature and salinity changes within a water mass are gradual.

If one mass is mixing with another, above or below it, the TS points in the zone where mixing is taking place fall on a straight line joining portions of the curve relating to the two different masses. The masses themselves are represented by the "corners" in the curve or by its end points (see Figures 5.1 and 5.2). Temperature and salinity changes between water masses are often abrupt.

It is assumed that all the water in a water mass has a similar origin and history (von Arx, 1962). A water mass is "conditioned" (acquires its properties) at the surface, while it is in contact with the atmosphere. That is, it absorbs solar heat while at the surface, and acquires its characteristic salinity which depends on evaporation rate and rainfall, and to a lesser extent on the melting and freezing of ice, and runoff from the land. Thereafter it conserves the characteristics it has acquired even though it moves. Cold polar water, for instance, being of high density, sinks and slides toward the equator on or near the bottom. Thus water masses form tilted layers of enormous areal extent. Those at the surface at high latitudes are encountered at ever-increasing depths as the equator is approached. For example, at the surface in far-southern latitudes, Circumpolar Water, and to the south of it, Antarctic Bottom Water, form concentric zones around the Antarctic continent. In tropical latitudes, these water masses are encountered at great depths, with Circumpolar Water forming a layer overlying Antarctic Bottom Water (they are labeled R and S respectively on the solid curve in Figure 5.2).

The water masses are in constant movement, both vertically and horizontally. Climatic zonation is the cause of this movement, and if it were not for the perpetual circulation of the water masses, deep water would never be renewed and reoxygenated.

It should be noticed that the water in the ocean's uppermost 200 m is not regarded as belonging to water masses since conditions in it are not conserved but vary in response to extrinsic causes such as sun, wind, and surface currents. This shallow surface layer of the oceans, to which belong, also, the waters of the neritic (shelf) seas, is therefore, by definition, excluded from the water masses.

The constant circulation of the water masses, which ensures the periodic return of deep water to the surface, is thus necessary for the continuance of life at great depths. But, as remarked above, it is uncertain to what extent a water mass can be regarded as a "region" with its own fauna. Funnell (1971) considers that faunal change at water mass boundaries, like temperature and salinity changes, is abrupt.

Berger (1969) has explored the problem as it affects the planktic* foraminifera, with a view to judging whether occurrences of foraminifera species are correlated with conditions in the surface layer or with conditions in the water mass below it. He reasoned that a species' "home" water should be less variable in temperature and salinity than water in which it chances to occur by accident. Therefore if a foram species is controlled by conditions in a water mass, the temperatures and salinities of samples containing the species should be less variable in "deep" samples (from 150 m) than in "shallow" samples (from 10 m). The results were inconclusive, but an interesting complication has also to be considered. It had been shown by Ericson and Wollin (1962) that for some foram species, specimens in surface waters (in the photic zone, at depths of less than 200 m) are preponderantly thin-walled, immature forms, whereas specimens from greater depths are preponderantly thick-walled, reproductively mature forms. This raises the question: is a species' distribution governed by the tolerance limits of its mature or immature forms? The question is one of general interest in a wide array of contexts.

Benthic Faunas

Benthic faunas are obviously classifiable on the basis of depth. Such a classification can be thought of as ecological. Two of the biomes in a coarse classification of the world's biosphere are the shallow sea benthos and the deep sea benthos (page 16). However, in the marine context it is artificial to treat ecological and taxonomic classifications separately. Therefore, we here consider the characteristics of the depth zones of the benthos and how they may best be defined.

Numerous classifications have been proposed. Several of them have been described and compared by Menzies, George, and Rowe (1973). These authors stress that within a given geographic area there are often fairly abrupt discontinuities in the composition of the benthic fauna at certain particular depths, so that these depth contours constitute natural boundaries for a zonal classification. However, the absolute depths of these faunal changeovers vary with latitude, and from one ocean to another. Therefore, the zoological distinctness of the zones is lost if data

---

*The adjectives *planktic* and *benthic* are better constructed etymologically, and shorter, than *planktonic* and *benthonic*. See Martinsson (1975).

from all oceans are pooled before a zonal classification is attempted. In constructing their classification, Menzies and co-workers first recognized the zones in each of a number of restricted geographic areas, and subsequently linked up the boundaries they had recognized between one area and another.

It is important to keep in mind that the boundaries of the faunal zones, characterized taxonomically, need not correspond with the boundaries of the geomorphological zones—intertidal, continental shelf, continental slope, continental rise, abyssal plains—characterized topographically. Quite often they do correspond, but not always. There is no *a priori* reason why they should. When discussing the zonation of the sea floor, therefore, it is important always to be clear as to whether topographical or biological criteria are under consideration.

The four major faunal zones* that Menzies and co-workers (1973) recognize are as follows: the Intertidal Zone, the Shelf Zone, the Archibenthal Zone, and the Abyssal Zone.

The Intertidal Zone is absent from the High Arctic and the High Antarctic because their shores are scoured by ice. In these latitudes some typically intertidal genera occur subtidally, in shallow waters of the Shelf Zone.

The Shelf Zone is faunally distinctive but the depth of its lower boundary (which is determined, of course, by faunal composition, not by topography) varies enormously, from less than 10 m at some places in the Arctic, to more than 1200 m off the coast of Peru. It is divisible into two subzones, photic and aphotic, according to whether light penetration is sufficient or insufficient for sessile algae to grow. The depth of the boundary between the subzones depends on the local transparency of the water.

The Archibenthal Zone (or Archibenthal Zone of Transition) is entirely aphotic. It is transitional between the Shelf and Abyssal Zones in temperature and sediment type, as well as in fauna, since in most latitudes its topographical position is on the continental slope. The continental slope is very much a transition zone in the geomorphological sense; it forms the transition between the continental and oceanic segments of the earth's crust, between surfaces that are subaerial at least occasionally and surfaces that are perpetually submerged (page 28). This greatly influences the nature of the sediments. On the shelf the sediments are largely terrigenous; in the abyss they are formed of clays and pelagic oozes, the accumulated remains of dead planktic organisms, chiefly foraminifera,

*Menzies, George, and Rowe (1973) call these zones *Provinces* and use the term *zone* for subdivisions of their provinces. Here, only the larger units (their Provinces) are described and are called *Zones* for clarity.

diatoms, radiolaria, and pteropods. The changeover occurs on the continental slope and a range of sediment types is thus included in the Archibenthal Zone. The Archibenthal Zone is more or less equivalent to the Bathyal Zone of many other authors.

The Abyssal Zone embraces the fauna of all the rest of the sea floor. Its salient physical characteristic is that the range of variation of water temperature in this zone is everywhere less than 2°C. The constancy of the temperature, rather than its absolute value, appears to be the important factor. Therefore the physically defined line that corresponds most closely with the upper boundary of the faunal zone is the isopleth joining points whose *range* of temperatures is 2°C; this isopleth does not, of course, coincide with any particular isotherm.

The four zones have distinct faunas. The degree of zone-endemism differs from zone to zone and is least in the Archibenthal (Transition) Zone. (The degree of species endemism of a zone is the proportion of the species occurring in that zone that occur nowhere else; and similarly for generic endemism.)

The zones are very much more distinct from each other at low than at high latitudes. That is, zone-endemism is much more pronounced at low latitudes. At high latitudes the zones are telescoped together. Thus in the far north, abyssal faunas occur in waters so shallow that, topographically, they are on the deeper part of the continental shelf. This telescoping of the zones is illustrated diagrammatically in Figure 5.3. If the range of variation of water temperature is the chief factor determining the limits of faunal zones, the telescoping at high latitudes is not surprising. In polar seas, the water is always cold, and always at a temperature not very different from that at abyssal depths everywhere in the ocean. By contrast, at low latitudes, although temperatures at depth undergo only slight seasonal variation—and the deeper the slighter until all variation vanishes—shallow water temperatures vary markedly.

The discussion in the preceding paragraphs relates to the depth zones of whole faunas. Individual taxa—genera and species—all have their own depth ranges, which are not necessarily coincident with those of the faunas to which they belong; some have narrower ranges, and some— those that are not endemic to a zone—have wider ranges. Species and genera may be arbitrarily classed as *stenobathic* or *eurybathic* depending on whether their depth range is less than or greater than 300 m. The upper and lower limits of a taxon's zone depend on a number of imperfectly correlated factors, depth itself, temperature, range of temperature variation, salinity, currents, sediment type, and, for organisms that live in the photic zone, light penetration. Thus the position and width of a taxon's zone vary from one geographic location to another.

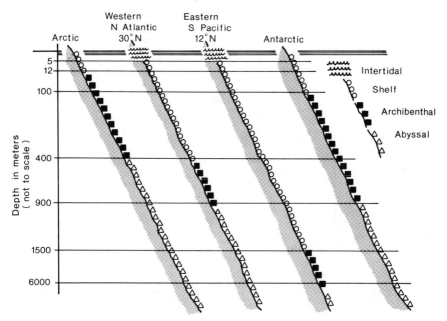

Figure 5.3   The approximate depths of the boundaries of the four benthic faunal zones at four different geographical locations. The gradient of the sea floor has been drawn with an (apparent) uniform slope to emphasize that faunal zones are defined independently of topographic features. Note that the depths are not to scale. (Data from Table 9-3 of Menzies, George, and Rowe, 1973.)

The variation with depth of the species-richness of the benthos has been the subject of much inquiry. Vinogradova (1962) presented graphs showing how the numbers of species in certain groups vary with depth; (she considered only those species known to occur below 2000 m). Data from many geographic locations were pooled. One of her graphs, that relating to all species combined (a total of 1144), is given in modified form in Figure 5.4. The groups concerned are the Porifera, Coelenterata, Crustacea (Cirripedia, Isopoda, and Decapoda), Pycnigonida, Echinodermata (Crinoidea, Asteroidea, Echinoidea, and Holothurioidea), and Pogonophora. As may be seen, the number of species falls off rapidly with increasing depth down to about 6000 m and thereafter remains low and fairly stable down to the bottom of the deepest ocean trench. Some authors (for example, Wolff, 1970) treat depths below 6000 m, which include the trenches, as a distinct faunal zone, the Hadal Zone. The fauna of this zone is poor in species and a large proportion of them are endemic to local regions and to the different trenches.

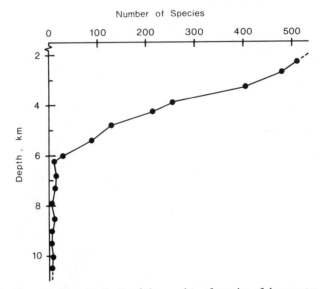

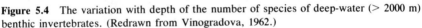

Figure 5.4   The variation with depth of the number of species of deep-water (> 2000 m) benthic invertebrates. (Redrawn from Vinogradova, 1962.)

There has been considerable disagreement over whether the abyssal fauna is cosmopolitan or regional.

One opinion, as given by Sylvester-Bradley (1971), is that there "seems to be a worldwide *psychrospheric* fauna common to the deeps of all oceans."* It is argued that the floor of the world ocean is a unit, not subdivided by barriers, and is the home of a homogeneous abyssal benthic fauna.

The converse opinion has been advanced by Vinogradova (1959) and by Menzies and co-workers (1973). They show that the species with the greatest geographic ranges are those of middle depths, those whose depth zones correspond, approximately, with the lower levels of the continental slope. In terms of the faunal zones, the proportion of geographically wide-ranging species is greatest among the members of the Archibenthal Zone fauna. Going upward from this zone, the species of the Shelf Zone and of the Intertidal Zone contain progressively smaller percentages of wide-ranging species. Going downward from the Archibenthal Zone, the percentage of wide-ranging species again falls off. As examples of wide-ranging taxa of the Archibenthal Zone, Menzies and co-workers list 11 species (3 sponges, 3 coelenterates, 3 echinoderms, 1 annelid, and 1

---

*From the Greek "psychros," cold. The psychrosphere is the continuous, cold, bottom layer of the world ocean.

crustacean) that occur on the continental slope on both sides of the North Atlantic. Another 15 species of the eastern North Atlantic have very closely related species on the western side. Possibly the most wide-ranging of all marine organisms are the sea-pens (sessile coelenterates) of the *Umbellula* species complex.

The extent of a species' geographic range depends on its bathymetric (depth) range as well as on the depth of the midpoint of its zone. As one would expect, the more eurybathic a species is, the greater its geographic range tends to be. Conversely, stenobathic species have small geographic ranges. Vinogradova (1959) summarized available data on "abyssal" species (those found below 2000 m) as follows. The percentage of eurybathic species is:

0% of species with very restricted geographic range;
30% of species whose range covers half an ocean;
65% of species whose range coincides with one ocean;
85% of species whose range includes two oceans;
100% of species with a "pan-oceanic" distribution.

These results suggest that the "abyss" is not a unit. Species at the greatest depths are confined to basins surrounded by submarine mountains, hills, and ridges. The lowest parts of the abyssal plains, as well as the deep ocean trenches, are to the marine world what islands are to the terrestrial, regions in which organisms can evolve in isolation, and between which they cannot easily disperse.

The same is true of species at shallow depths, in the intertidal and on the shelf. The shelf areas of the world do not form an unbroken whole.

It seems a reasonable speculation that those species with the greatest geographic ranges are the ones occupying a zone that, looked at on a map, forms a connected network of worldwide extent or, if no such network exists, then one approaching it as closely as possible. On a world map showing the bathymetric contours of the oceans, it is easy to search for a depth zone with the property that any point in it is accessible from any other without straying outside the zone. Such a depth zone would not necessarily meet our present requirement, however. As remarked above, the faunal zones do not correspond with physically defined depth zones. A given faunal zone occupies progressively shallower depths at higher and higher latitudes (see Figure 5.3). This fact needs to be allowed for when one searches for the zone of greatest connectedness available to benthic organisms. In this zone, rather than at greater depths, the proportion of cosmopolitan (pan-oceanic) species might be expected to be greatest. But such a zone would not have maintained a constant position

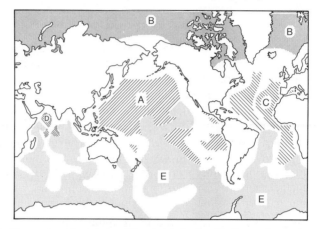

Figure 5.5 Zoogeographic regions of the abyss. A classification and mapping proposed by Menzies, George, and Rowe. The regions are the Pacific (A), Arctic (B), Atlantic (C), Indian (D), and Antarctic (E). (Adapted from Menzies, George, and Rowe, 1973.)

through geological time. Its depth must have varied in response to climatic trends; and continental drift must have wrought changes in its horizontal pattern by causing some oceans to expand and others to contract.

Since the lowest depths of the oceans are partitioned into regions—basins and trenches—by topographic barriers, a zoogeographic classification of the abyss is possible; the abyss should perhaps not be regarded as a single, indivisible psychrospheric faunal region. A classification proposed by Menzies and co-workers (1973) recognizes five regions, the Arctic, Pacific, Atlantic, Indian, and Antarctic Deep-Water Regions (see Figure 5.5). They subdivide the regions into a total of 13 provinces, some of which are further subdivided.

2. CHANGING DISPERSAL ROUTES AND CHANGING MARINE CLIMATES

Marine biogeography has been as profoundly influenced as terrestrial by continental drift, crustal warping, and long-term climatic changes. We consider them in turn, by examining particular examples.

The Effects of Changing Patterns of Land and Sea

Continental drift created and then destroyed the Tethys Sea. The Tethys existed first as an enormous embayment in the east side of Pangaea. It became the seaway separating Laurasia and Gondwana when Pangaea broke into two parts. For many millions of years it formed part of a circumglobal sea, probably with a westward-flowing surface current (Tedford, 1974). All that remains of it now is the modern Mediterranean and there is considerable disagreement over the details of its history (see page 33). Fossil evidence suggests that its existence as a continuous seaway ended at the time of the Oligocene-Miocene transition when, as a result of plate drift, Africa-Arabia came into contact with Eurasia. It appears that the marine faunas of the Indian and Atlantic (including Mediterranean) Oceans began to differentiate at this time; and that, simultaneously, the migration of land mammals (including elephants) between Asia and Africa resumed. Thus evidence from both marine and terrestrial fossils points to the same conclusion (Middlemiss and Rawson, 1971).

Another part of the world where the formation of a land bridge, and the consequent closing of a seaway, has been of great biogeographic importance, is Panama. As already discussed (Chapter 2, page 34), the Isthmus of Panama formed in the Pliocene, about 6 m.y. BP according to Raven and Axelrod (1974). As a land bridge, it permitted the migration and intermingling of North American and South American land mammals and the consequent extinction, presumably because of competitive exclusion, of many species. Its effect as a barrier to the dispersal of marine faunas has also been pronounced. A previously homogeneous marine fauna has become divided into Pacific and Caribbean parts, in which "geminate" species, that is, poorly differentiated pairs of species descended from a common ancestor, have developed (Middlemiss and Rawson, 1971). Another outcome of the evolutionary divergence of the western Atlantic and eastern Pacific faunas is that now the fish faunas have only 1% of their species in common (van den Hoek, 1975). Extinctions have occurred also. According to Briggs (1974), after the continuous Central American sea became separated into two parts by the formation of the isthmus, 43 genera and subgenera of mollusks that had belonged to the combined fauna went extinct on the Atlantic side, and 4 on the Pacific side. A possible reason for the contrast between these two numbers is discussed below (see page 164).

A third region where a seaway has alternately appeared and disappeared, as a land bridge sank and rose, is, of course, Beringia. The biogeographic effects, both on the land biotas of the Palearctic Region, and on the marine biotas of the North Pacific and Atlantic-Arctic Oceans, were discussed in Chapter 4 (page 130).

The preceding paragraphs have mentioned three geographic regions (Tethys, Panama, Beringia) where dispersal routes have undergone a "make-and-break"; that is, where the emergence (or submergence) of a land bridge has amounted to the breaking (or creation) of a seaway. In all cases there has been a *convergence* of the joined biotas and a *divergence* of the separated biotas.

Additional examples of converging and diverging biotas have been given by Hallam (1973). Thus the mammal faunas of Europe and Asia appear to have converged in the Eocene, probably because of the drying up of the Turgai Strait, the epicontinental sea that had previously joined the Arctic Ocean and the Tethys Sea.

The effect of continental drift in simultaneously narrowing the Pacific and widening the Atlantic has also caused biotic convergences and divergences. Thus the bivalve mollusk fauna of the American and Asian shores of the Pacific have converged biotically as they have approached each other spatially. At the same time, the bivalve mollusks, and also the benthic foraminifera, of the Caribbean and Mediterranean have diverged as the Atlantic grew wider. Not only has the distance between these two shelf seas increased but, at the same time, the shallow-water dispersal route that once linked them, around the northern end of the Atlantic, has become too cold to be usable (Hallam, 1973).

It is important to stress the *un*symmetry of biotic convergence and divergence. They are entirely dissimilar processes. When biotas come into contact and converge, individual organisms of many previously separated species-populations, and their descendants, begin to mingle. They disperse into each others' territories and, if they are on the same trophic level, compete with each other; those on different trophic levels establish new predator-prey relationships. And related species, if they are interfertile, form hybrids.

When biotas are split and diverge, the differentiation that takes place is a purely evolutionary process. It seems to be commonly assumed that the divergence is gradual and that when it has been in progress for only a short period evolutionarily speaking, the amount of differentiation will be slight and closely similar sister-pairs of species ("geminate pairs") are to be expected; examples where they have indeed been found are the bivalve mollusks on either side of the Isthmus of Panama (see above, page 151), the paired species of invertebrates of the western and eastern North Atlantic segments of the Archibenthal Faunal Zone (page 149), many of the large mammals (page 135), and angiosperm and gymnosperm genera (pages 132 and 133) of temperate North America and Asia, and also of North America and Europe (see page 206).

These gradually diverging taxa constitute examples of phyletic gradualism in operation (see page 67), and apparently demonstrate that it does indeed happen. If rectangular speciation—the abrupt appearance of large differences between populations that have become reproductively isolated—is occurring too, evidence of it seems not to have been sought for or noticed.

A consequence of continental drift unrelated to the making and breaking of dispersal routes is the fact that continents and their margins have drifted across zones of latitude and hence of climate. For example, in the Jurassic the east coast of North America was tropical. It trended east-west and lay at about 25°N (Dietz and Holden, 1970; and see Figure 2.1). Coral reefs grew in the shelf zone as far east along the coast as the location of the modern Grand Banks of Newfoundland, now at a latitude of 45°N and well outside the climatic zone in which reef corals can grow given the present world climate.

Climatic change has also been taking place on a worldwide scale, independently of the drift of the continents, and we next explore its effects on marine biotas.

The Effects of Worldwide Climatic Change

As the earth's climate has changed through geological time, the physical properties and locations of the water masses, and their associated biotas, have presumably changed too. Little is yet known, however, about how these changes are interrelated. In particular, it is not known for how long existing water masses have had their present separate identities and boundaries. The TS curves characterizing the major modern water masses have shown no change since they were first observed 50 years ago; this shows that they are not ephemeral (Neumann and Pierson, 1966; McGowan, 1971). It is conceivable, therefore, that they are more or less permanent structures, retaining their individuality even while their temperature and salinity properties changed gradually, as they must have, through geological time. Funnell (1971) speculates on whether, as change took place, plankton organisms migrated from one water mass to another in order to remain in suitable environments, or whether they stayed in their home water masses and evolved synchronously so as to remain adapted to the changing conditions.

The answer to this question is obviously of crucial importance in deciding whether or not one can correctly infer past marine climates from the assemblages of microfossils found in sediment samples. For suppose marine microorganisms remained morphologically unchanged while undergoing continuous physiological adaptation that kept their requirements always matched to the physical environment in which they found themselves; if this were to happen, it would clearly be impossible to infer paleoclimates from microfossils as is so confidently done.

The plankton organisms most used in attempted reconstructions of past conditions are the Foraminifera or "forams."[*] They are Protozoa of the class Sarcodina (Rhizopoda) and are, in effect, amoebas with shells. The shells (tests) have exceedingly intricate species-specific patterns, and accumulate without decaying on the sea floor. Their abundance, distinctiveness, and sensitivity to environmental conditions make them "indicator fossils" par excellence. Before discussing them in more detail, other members of the plankton should be mentioned.

Johnson and Brinton (1963) discussed the geographic ranges of metazoan species of the holoplankton,[†] such organisms as chaetognaths, euphausids, pteropods, and copepods. They found that epipelagic and mesopelagic species responded to the temperatures of the waters in which they lived and thus had different patterns of geographic ranges. Species living in the epipelagic zone, that is, in surface waters in contact with and affected by the atmosphere, have geographic ranges determined by surface temperatures. Likewise for many mesopelagic species, the boundaries of their ranges coincide with isotherms at the depths they inhabit. Such an observation is not unexpected, but is has interesting implications. As we shall see below, the climatic history of the surface layers of the ocean and of the underlying water masses may have been controlled by quite different, and unrelated, events. If this is indeed so, it means that epipelagic and mesopelagic species, even though they may be closely related taxonomically, occupy environments that have had very different histories.

The plankton species that are most valuable as indicators of past marine climates are those with fossilizable skeletons (Funnell, 1971). They belong to six main groups: the planktic foraminifera and coccoliths (with calcite shells), the diatoms, radiolarians and silicoflagellates (with skeletons of opaline silica), and the pteropods (with shells of aragonite which, like calcite, consists of calcium carbonate, but which has crystals of different form).

The most important of these, as already remarked, are the forams, but their fossils persist in the sediment only if the bottom is shallower than the $CaCO_3$ compensation depth, which varies somewhat from place to place in the range 4000 to 5500 m approximately (Murray, 1973). The concentration of carbon dioxide in the sea increases with depth and the compensation depth is the depth at which the concentration becomes sufficiently

---

[*]Forams are abundant in the benthos as well as in the plankton. Indeed, the number of benthic species is enormously greater than the number of planktic species. The present discussion refers only to planktic forams, however.

[†]The species of the *holoplankton* (permanent plankton) are planktic throughout their lives. The *meroplankton* (temporary plankton), in contrast, consists of planktic larval forms (for example, of fish and mollusks) whose mature stages are members of the nekton or benthos.

high to dissolve calcareous shells. Foram tests are therefore absent from the deep abyssal plains although in shallow waters they are often the commonest of all microfossils.

Living planktic forams occur chiefly in the uppermost 100 m of the sea and chiefly in oceanic, as opposed to near-shore, waters. Samples of living planktic forams, obtained in plankton tows, show them to be distributed very irregularly, as are all forms of plankton; patches with inexplicably high density occur, both horizontally and vertically. Even so, they appear to form distinct faunas, each characteristic of a particular surface water mass (Phleger, 1960). This suggests that among species in which the immature forms occur at the surface, and reproductively mature forms in deeper water, it is the former that control the species' geographical ranges (see page 140).

"Dead" planktic forams that settle on the sea floor are the empty tests left by reproducing individuals. When reproduction, sexual or asexual, takes place, the whole of the parental protoplast divides into gametes or asexually formed embryos. The parental test is left empty, to sink to the bottom and add itself to the steadily accumulating sediment of tests. A sample core from such sediments therefore shows how the planktic community at that site, of forams and other fossilizable species, has varied through time. From this, the way in which marine climates have changed can be inferred.

Of the many possible sources of error, three are especially worth noting. First, the sediment at any place on the sea floor receives material from the whole overlying water column (supposing the tests fall vertically), and therefore it combines the input from all depths, perhaps from several distinct water masses stacked one on top of another; even so, it probably "reflects near-surface conditions, at least in large part" (Phleger, 1960). Second, as empty tests settle from the water, they may be displaced by currents. Also, they are not merely displaced, but to some extent sorted; tests of different species, and hence of different shapes and sizes, fall through the water at different speeds and this determines the distance the current can carry them before they reach the bottom. Third, species differ in preservability.

Notwithstanding these and other sources of error that must be allowed for, considerable progress has been made in inferring the history of marine climates. These inferences may be based on the presence of so-called hydrological indicator species and genera, such as certain species of the genera *Globigerina, Globorotalia,* and *Globeriginoides,* which are indicative, respectively of cold, medium, and warm water (Boltovskoy, 1969); and on the direction of coiling of the spirally coiled tests of the very abundant cold-water species *Globigerina pachyderma.*

Globigerina pachyderma is a species (one of many) with spirally coiled tests. Two varieties occur, a variety with left-coiling (sinistral) tests and a variety with right-coiling (dextral) tests. The species as a whole is restricted to cold waters; its abundance in the plankton increases poleward from temperate latitudes. At the same time, the proportion of sinistral individuals increases from less than 50% where its density is low, in comparatively warm water, to almost 100% where surface temperatures are near 0°C (see Figure 5.6). The proportions of sinistral specimens of the species in samples of fossil communities (thanatocoenoses) obtained from sediment cores therefore serve as indicators of contemporary water temperatures. For example, high proportions of sinistral specimens at relatively low latitudes in the late Tertiary suggest that temperatures were lower then than now (Kennett, 1968).

Climatic fluctuations in the Pleistocene have also been investigated. The proportions of sinistral and dextral *G. pachyderma* have provided some of the evidence; other evidence has come from the species-composition of fossil communities of forams, and of radiolarians and diatoms (Hays, 1967). It is not clear whether changes in microfossil communities indicate the behavior of surface water temperatures which may have changed gradually and continuously in response to changes in atmospheric temperature; or whether the microfossils reflect the behavior of water masses. If they are responding to water masses, one would not necessarily expect clinal change in the composition of thanatocoenoses; stepwise changes would be more likely as the boundaries of water masses, moving as coherent entities, passed over a particular point on the sea floor. This may explain why Pleistocene communities of planktic forams very often show no changes corresponding to the glacial-interglacial climatic oscillations of the epoch. Perhaps evidence of these can be expected only in sediment cores taken at sites close to water mass boundaries, across which the boundaries were displaced as the climate changed (Funnell, 1971).

Microfossil assemblages of increasing age contain fewer and fewer extant species and inferences about climates earlier than the middle to late Tertiary have to be based on species now extinct, whose environmental requirements are, of course, unknowable. Even so, some inferences are possible. Forams are classifiable into "morphotypes" (which are not equivalent to genera) on the basis of their shapes, and different morphotypes are at present typical of different latitude belts. The proportions of different morphotypes at various latitudes and at various times therefore provide one way of inferring past climates.

In addition, the morphotypes represent different evolutionary stages; some are primitive, others evolutionarily advanced. Sudden evolutionary

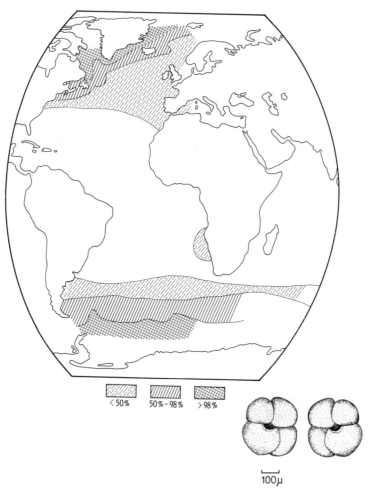

Figure 5.6 The percentage of left-coiling individuals in living populations of the planktic foram *Globigerina pachyderma*. Note that the species is restricted to cold waters. As right-coiling becomes more frequent, toward lower latitudes, the abundance of the species decreases; it is not found where surface temperatures exceed 24°C. (Adapted from Bé and Tolderlund, 1971.) Lower right: Sinistral and dextral specimens (on left and right respectively) of *G. pachyderma*.

radiations, which produced numbers of new forms, and sudden waves of extinctions, which caused abrupt declines in diversity, have been attributed to radical changes in marine climates. On such evidence, Cifelli (1969) has suggested that twice in the past—in the late Cretaceous and again in the middle Tertiary—ocean waters may possibly have been uniformly cool. No mere shifting of water masses toward the equator is envisaged, but a weakening of the latitudinal temperature gradient, so that the oceans were more homogeneous, as well as cooler, than they are today. A warm period in the Eocene, between the cool periods, is suggested by the evolutionary diversification of morphotypes that occurred, and also by the presence of presumed warm-water forms in high latitudes (Funnell, 1971).

It was remarked above that the spirally coiled tests of *Globigerina pachyderma* may be sinistrally or dextrally coiled and that the proportions of these two kinds vary geographically (see Figure 5.6). Bolli (1971) has discussed the topic of coiling direction in several species, in some detail. He considers that in primitive species, dextral and sinistral individuals are equally common; evolutionary advance is accompanied by the development of a steadily more pronounced preference for one coiling direction or the other. In the most advanced species, 95% or more of the individuals in a population coil the same way. The preferred direction may be the same worldwide; or (as in *G. pachyderma*) it may be mainly dextral in some populations and mainly sinistral in others, and hence vary geographically. The way in which the preferred coiling direction in advanced species varies with time can be observed in sediment cores. It is found that in some species the preferred coiling direction may change suddenly (in the geological sense) from dextral to sinistral or vice versa. For example, cores from the Caribbean show that *Globorotalia menardii* has switched coiling directions, abruptly, 14 times since the early Miocene. These abrupt alternations are nearly always from one clear direction preference to the opposite; in only a single, exceptional species (of those studied) was there a reversion to the primitive state with dextrals and sinistrals about equally common. Clear interrelationships between the spatial and temporal variations in preferred coiling direction should reveal much concerning the marine biogeography of the past.

The foregoing paragraphs have dealt solely with planktic organisms and their fossils, and the light they shed on present and past surface water climates. The geological history of the deeper layers of the ocean is probably quite different from that of the overlying water. Organisms of the deep-sea benthos are obviously the appropriate indicator species to use in attempting to unravel the course of events in the abyss; the remains of shallow water plankton that have merely drifted down through this water are uninformative.

Benson (1975) has investigated 80 species of fossil ostracods in sediment cores from 77 sites in the deep sea. Ostracods are shelled microcrustaceans of the benthos and are common in both fresh and salt water. The shells (carapaces) of shallow-water species are rather smooth and featureless but those of deep-sea species are morphologically much more complex. As with forams, it seems reasonable to assume that the correlation between carapace structure and water temperature observed in modern species held also in the past; if so, past conditions can be inferred from the "morphotypes" of extinct fossil ostracods as well as from the known environmental preferences of still extant species.

According to Benson, the 8°C isotherm on the sea floor represents, approximately, the dividing line between two very distinct ostracod faunas, a warm-water fauna and a cold-water fauna. The 8°C isothermal surface in the sea separates the so-called "warm-water sphere" (thermosphere) and "cold-water sphere" (psychrosphere) of the ocean, that have been likened, respectively, to the troposphere and stratosphere of the atmosphere (see Neumann and Pierson, 1966). At present the depth of the boundary surface is at about 700 to 800m in low latitudes and becomes steadily shallower toward the poles. The Abyssal Faunal Zone of the benthos is entirely within the psychrosphere (see page 148) and so also is much of the Archibenthal Faunal Zone. The higher faunal zones are in the thermosphere, which covers the continental shelves and, in the deep sea, floats on top of the denser, colder psychrosphere.

Arguing from the evidence provided by fossil ostracod assemblages which are assumed to represent thermospheric and psychrospheric faunas, Benson (1975) has inferred that the modern division of the ocean into two contrasted layers has not been permanent. The fossil evidence suggests that throughout the early Tertiary the whole ocean was warm and was not divided, as it is now, into warm and cold "spheres." The psychrosphere came into existence suddenly (geologically speaking) in the Eocene, about 40 m.y. BP. Possible causes of this event were the closing of the Tethys Sea, which had supplied large amounts of warm water to the world ocean, and the availability, as a result of continental drift, of the two polar oceans as sources of cold water.

The ostracods now living in the psychrosphere are primitive Mesozoic groups and are presumed to have been able to adapt to the drop in temperature that occurred when the psychrosphere came into existence. More advanced ostracod families, that have evolved within the thermosphere since the Eocene, have not been able to migrate downward to greater depths because the formation of the psychrosphere has "closed much of the world-ocean floor to further invasion and forced those species already there towards extreme modes of adaptation" (Benson, 1975).

It is not clear how (or whether) this theory of the abrupt formation of the psychrosphere in the Eocene meshes with Cifelli's (1969) theory that, twice in the past, ocean waters have been uniformly cool (see page 157).

To end this section, it is worth recalling some of the points that have been touched on in the last few pages. We have considered the pelagic and the benthic faunal realms; planktic and benthic organisms, living and fossil; forams and ostracods; extant and extinct species, the environments of the latter being inferred from their morphotypes; dextrally and sinistrally coiled forams and their use in interpreting past marine climates; the separation of the ocean into a thermosphere and a psychrosphere; the possibility that the oceans may at times have been more homogeneous (less layered) than at present; and the possible evolutionary effects of changes in the ocean environment. This list emphasizes the variety of different kinds of evidence that have been used to infer the history of the oceans and emphasizes also the extraordinary complexity of the task. It will be noticed that the motive for much of the work has been to advance geological and oceanographic knowledge. The fact that living or once-living organisms are the chief source of evidence is often treated as incidental. It is time for biogeographers to turn many of the arguments around, and to examine the earth scientists' conclusions from the biologists' point of view, for the light they throw on evolutionary biogeography.

3. LATITUDINAL ZONATION, PRESENT AND PAST

It was remarked earlier (page 15) that latitudinal zonation is more apparent in the geographic ranges of marine than of terrestrial organisms and biotas. It was also remarked (pages 19 and 97) that littoral organisms have one-dimensional ranges, representable on a map by line segments. This suggests that the biogeography of littoral organisms along north-south trending coastlines should be especially revealing. Shorelines are a useful testing ground for all manner of theories about such matters as the latitudinal diversity gradient, evolutionary divergence as a result of range fragmentation, dispersal mechanisms and rates, and the like. It is a fortunate accident of geography that the world has many long north-south trending coastlines available for study. The most thoroughly studied of all is probably the east coast of North America, and the west coast runs it a close second, so in much of what follows attention will be primarily focused on these two coasts. It is also fortunate that many of the most characteristic and abundant members of littoral communities are mollusks, whose shells form well-preserved and easily recognizable fossils.

   We begin with a consideration of the subdivision of modern coastlines into natural biotic provinces (the word *province* is used here without any connotation of rank, to mean a subdivision of any hierarchical level). The evidence seems overwhelming that the boundaries of biotic provinces are determined by modern abiotic (that is, physical) factors. Thus Hayden and Dolan (1976) classified the Pacific and Atlantic coastlines of North and South America twice, using biotic criteria for one classification and abiotic criteria for the other. Their abiotic provinces were classified on the basis of temperature, salinity, major currents, and the longshore currents very close to the shore; their biotic provinces on the distributions of 968 species of ascidians, crabs, and mollusks; (and see Chapter 1, page 22). They found a strong tendency for the biotically and abiotically defined provinces to coincide.

This leaves uncertain the reason why provinces remain distinct. One of two possibilities is that each offers unique environmental conditions, to which species from other provinces are unadapted; intruders therefore cannot establish themselves in a "wrong" province although nothing prevents their entering it. The other possibility is that actual barriers to dispersal exist that are difficult to cross.

The existence of many of the major boundaries now recognized on American shores could be satisfactorily accounted for by either theory. For example, consider the Atlantic coast of North America from the Bay of Fundy to Florida. There are three major biotic boundaries, at Cape Cod, Cape Hatteras, and Cape Canaveral (see Figure 5.7). They are the boundaries of mollusk provinces (Abbott, 1968; Hecht, 1969), algal provinces (Humm, 1969; van den Hoek, 1975), and invertebrate provinces in general (Hutchins, 1947). They separate regions of different environmental character, with different marine climates. At the two southern capes there is a very rapid change in surface water temperature, as shown by the bunching of the isotherms. Also, the capes themselves are barriers. They deflect currents and may hinder the easy transport of propagules (spores, eggs, larvae, algal fragments, adult animals) along the shore. And Cape Hatteras interposes a stretch of shore that is extremely inhospitable to all but a few littoral species; the long sandy beaches provide no suitable ground for attached algae, and in the shallow lagoons behind the outer beaches the waters are turbid and subject to pronounced temperature and salinity fluctuations (van den Hoek, 1975).

Where such barriers to dispersal coincide with boundaries between differing environments, it is difficult to judge the relative importance of the two factors in maintaining the distinctness of biotic provinces. They undoubtedly reinforce each other.

Now consider what has been happening over "geological" stretches of time. As climate has changed, water masses have shifted alternately northward and southward; as well, the steepness of the latitudinal gradients in physical factors of the environment, most importantly temperature, has been alternately large and small. Simultaneously, biotas have

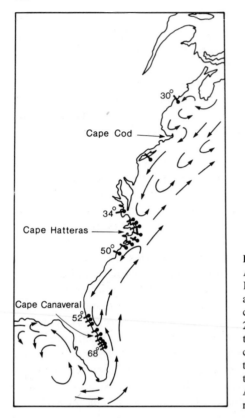

Figure 5.7 The Atlantic coast of North America from the Bay of Fundy to Florida showing currents near the shore and the points at which winter isotherms cut the coast. The isotherms, in °F and at 2°F intervals, show minimum sea surface temperatures in February. Observe how closely they are grouped at Cape Hatteras and Cape Canaveral. (Data from the Oceanographic Atlas of the North Atlantic Ocean, U.S. Naval Oceanographic Office, Washington, D.C., 1965.)

shifted to keep pace with shifting environments; and while this has been happening, all species have been evolving, at rates that no doubt differ markedly from one phylum to another.

For instance, Hecht (1969) has presented evidence on the distribution of 168 Miocene mollusk species suggesting that, in the Miocene, there was no faunal discontinuity, as there is now, at Cape Hatteras. The data are "somewhat problematic" because of different treatment of fossil material by different taxonomists, and because Miocene fossils are easily accessible only between Florida and New Jersey. However, if it is reasonable to infer patterns of marine climate from those of contemporary mollusk faunas, then the conclusion that the Miocene pattern differed markedly from the modern pattern seems acceptable.

This raises the problem as to whether there is any appreciable lag in the response of individual species and whole biotas to changing and shifting marine climates; and if there have been lags, their durations. In other words, should the present-day biogeographic patterns of littoral biotas be attributed to modern or to historical causes? It is a problem that has not been squarely faced.

In favor of the hypothesis that there is no lag in the response of biogeographic patterns to worldwide changes in abiotic factors and their gradients is the fact that there appear at present to be no conspicuous mismatches between biotas and their environments. This may, of course, be an observational artifact. Investigators are apt to take for granted that biotas are adapted to the environments in which they are found and to *define* an environment as being appropriate for a biota (or a biota for an environment) because the one contains (or is contained in) the other. More, it is probable that interesting anomalies often go unreported simply because the person who perceives them dismisses awkward facts merely as meaningless "noise" in the data.

A fact that supports the alternative hypothesis, namely that changing biogeographic patterns lag appreciably behind changes in the abiotic environment, is that in extant benthic foraminifera (Durazzi and Stehli, 1972) and in extant bivalve mollusks (Hecht and Agan, 1972) tropical genera tend to be the youngest, geologically; the average age (in millions of years) of the genera in each latitude belt increases with increasing latitude (see Figure 5.8); (the age of a genus is the age of its oldest known fossil member). The correlation of age with latitude is conspicuous. Durazzi and Stehli considered the average age of the genera of extant benthic forams in samples from 104 sites scattered between 50°S and 80°N latitude; the average age within a sample ranged from 40 m.y. (Eocene) back to 220 m.y. (early Triassic). Hecht and Agan considered the genera

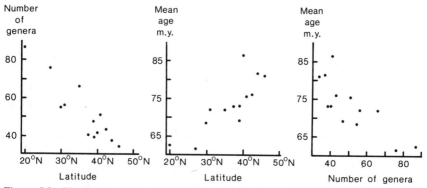

Figure 5.8 The three pairwise relations, among latitude, number of genera, and mean age, of 13 living bivalve faunas of the Atlantic coast of North America. The mean age of a fauna is the mean of the geological ages (ages of the oldest known fossils) of the taxa comprising it, in this case genera. (Data from Hecht and Agan, 1972.)

of extant bivalve mollusks in 13 latitude zones along the Atlantic shore from the Caribbean to New Brunswick and found average generic ages ranging from 61.6 m.y. (Paleocene) to 86.5 m.y. (Cretaceous). These authors conclude that speciation is more rapid in the tropics than elsewhere and that recently evolved genera have "not yet been distributed to higher latitudes." The implication is that, given time, they will be. It would be interesting to know whether evolutionarily more advanced genera are spreading into higher latitudes at a faster or slower pace than more primitive ones. This would throw light on a problem discussed in Chapter 3; see page 73.

Another piece of evidence showing that biogeographic patterns change slowly is the marked contrast between the North Pacific and the North Atlantic in the richness of their biotas, of both the shore and the ocean. The topic has been fully reviewed by Briggs (1974). Compared with that of the North Pacific, the North Atlantic biota (both flora and fauna) is conspicuously depauperate. The probable cause is that the North Atlantic was repeatedly and drastically cooled during the Pleistocene glaciations, while the Pacific was not. The physical contrast between the two oceans is believed to have arisen from the fact that no barrier blocked the flow of cold water from the Arctic into the Atlantic, but the Bering land bridge (which was in existence for much of the Pleistocene, see page 135) acted as a barrier protecting the Pacific. Thus North Atlantic surface water temperatures fluctuated widely and caused many species to go extinct. Further, biotic zones must have migrated alternately southward and northward. One result of these latitudinal shifts is that the littoral biota of the northern Gulf of Mexico closely resembles that in the same latitude in the Atlantic, although they are separated by the Florida peninsula whose southern tip now has a tropical biota; no doubt in cool periods, the biota of the modern warm-temperate Province (from Cape Hatteras to Cape Canaveral) was continuous all around the Florida peninsula. No comparable phenomenon appears to have taken place in the Pacific. The biotas on the outer coast of Baja California, and in the Gulf of California, are entirely different, and the presumption is that Pacific biotic zones have not migrated south and then north in a way that allowed species to migrate "around the corner" (Briggs, 1974).

Another contrast between the Atlantic and Pacific coasts of the Americas is that the geographic ranges ("spans") of individual littoral species tend to be much greater on the Atlantic side (see Figure 5.9). There are two possible causes for this, one in operation now, the other historical. The "modern" cause is that the "warm-water zone" with surface temperatures greater than 20°C is roughly twice as wide in the Atlantic as it is in the Pacific (Sverdrup, Johnson, and Fleming, 1942).

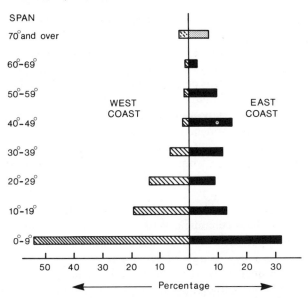

Figure 5.9 The "spans" (latitudinal extents of geographic ranges) of seaweed species (Rhodophyta, Phaeophyta, and Chlorophyta) on the west and east coasts of the Americas. The bars show percentages. The total numbers of species are: on the west, 1277; on the east, 684. (Data from Pielou, 1978a, b.)

How long this has been so is difficult to discover; in any case, it provides warm water species with much longer stretches of hospitable shore on the Atlantic side than on the Pacific side. The historic cause is that the Pleistocene extinctions were no doubt selective. *Stenotopic* species (those with narrow ecological tolerances) were presumably far more vulnerable than *eurytopic* species (with wide tolerances) to the temperature fluctuations; as a result, Atlantic biotas now contain unusually large proportions of eurytopes. It is interesting to observe that these extinctions have not obliterated the relation (shown in Figure 5.8) between the average geological age of bivalve mollusk genera and latitude. The implication is that ecological stenotopy and age are unrelated, or at least that there is no evidence to suggest that they are related.

The alternate northward and southward migrations of climatic zones would, as just remarked, affect species differentially in the sense that stenotopes would be at a selective disadvantage and hence more likely to become extinct. At the same time, this back-and-forth drift of climatic zones would tend actually to create eurytopes. To see this, envisage a warmth-loving species that migrates into high latitudes and establishes large populations there during a warm interglacial stage. Next, as a glacial stage sets in, the climate cools and this species' favored zone shifts equatorward again. However, among the large populations of the species left behind in the cold, there are likely to be some genotypes that can tolerate the reduced temperatures and can be the progenitors of local races or subspecies that will persist in high latitudes. The back-and-forth sweep of a particular set of environmental conditions will therefore, given time, spread an initially stenotopic littoral species over a very long stretch of coast. Natural selection acting on local populations will permit the differentiation and continued survival of morphs adapted to local conditions and they will differ from their ancestors. When several locally adapted morphs have been selected in this way, the species as a whole will be eurytopic.

This process requires that latitudinal shifts of climatic zones occur, as a necessary part of the mechanism. Without such shifts, there would be no reason for the appearance of large populations of a species outside its ancestral latitude zone. Large populations, as opposed to occasional strays, would be necessary to provide a sufficiently diversified array of genotypes, and hence sufficiently many potential survivors, when conditions deteriorated, to be the ancestors of a newly adapted subspecies.

The foregoing paragraphs discuss the contrast between the Atlantic and Pacific shores of the Americas in their ratios of eurytopes to stenotopes. This ratio also varies (at least among mollusks), on a small spatial scale, with depth on the shore. Jackson (1974) compared the extents of the geographic ranges of intertidal and subtidal mollusk species. The investigation was limited to tropical and subtropical representatives of four families of infaunal bivalves, the Lucinidae (lucinas), Cardiidae (cockles), Veneridae (Venus clams), and Tellinidae (tellins). "Intertidal" species— defined for present purposes as those living down to 1m below low water —can be classed as eurytopes since they must be able to endure widely fluctuating temperatures and salinities, and disturbance by storms; conversely, "subtidal" species are presumably relatively stenotopic. Jackson found that intertidal (eurytopic) species tended to have larger geographic ranges and also to be geologically older, on average, than subtidal (stenotopic) species. From these observations, he inferred that the intertidal eurytopes must be species that disperse easily. They would then be capable of colonizing all the sites suitable for them, which tend to be scattered at intervals along a long stretch of shore; they would not be endangered, as species, by frequent extinctions of purely local populations. Further, easy dispersal would ensure easy gene exchange and would delay divergence and speciation in local populations. The argument is that large geographical ranges and great geological age are two outcomes of a single cause. This accords with the observation (see page 164) that geologically young species are confined to smaller ranges than older species because they have not had time to expand their ranges. However, Jackson's (1974) argument does not stress the long-term, historical expansion of species' ranges, but attaches more weight to rapid dispersal among local populations.

This discussion of the latitudinal zonation of littoral biotas has yielded, as may be seen, a possible explanation of the present-day latitudinal gradient in species-richness or "diversity" (see page 90). It is that the ranges of many species that have recently evolved in the tropics have not yet had time to expand into high latitudes (Fischer, 1960). This may well be a contributory cause of the diversity gradient. The fact that there were latitudinal diversity gradients in the distant past—among brachiopod families in the Permian, and among planktic foram species in the Cretaceous (Stehli, Douglas, and Newell, 1969)—does not invalidate the argument; these past gradients, too, may have occurred soon after bursts of rapid evolutionary divergence in the tropics, while the newly formed species' ranges were still expanding.

This mechanism is probably not the only, or even the chief, cause of the latitudinal diversity gradient, however. Yet another possible contributory cause, applying to littoral mollusks at any rate, has been proposed by Taylor and Taylor (1977). They have shown that, in the North Atlantic, there is a large change in the number of taxa (species and families) of predatory gastropods at the latitude (about 40°N) where the primary productivity regime switches from continuous (throughout the year) to seasonal (summer only). The number of taxa of predatory gastropods south of this boundary greatly exceeds the number north of it. The change is particularly abrupt in the eastern North Atlantic. The probable explanation is that the continuous productivity south of the boundary assures an unvarying supply of a wide range of prey species and hence permits the successful coexistence of a large number of predator species with highly specialized diets.

It would be interesting to know how the latitude of this boundary has shifted as climate has changed. So long as the tilt of the earth's axis relative to its orbital plane around the sun has remained as it is now, a poleward limit to continuous year-round productivity must have been set by the number of hours of daylight per day, regardless of how warm it was. Much would become clear if we could only know how biotas were zoned latitudinally in a world with six months of winter darkness and six months of summer daylight in the polar regions (as now), but with a moderate climate everywhere and no polar ice (as in nonglacial periods).

The discussion of littoral biotas and their latitudinal zonation has also demonstrated how difficult it is to sort out the effects of modern and historic causes as they influence the biogeography of organisms now living. The boundaries of littoral biotic provinces appear to coincide with abiotic boundaries as they exist now. The contrast in diversity between the Atlantic and Pacific appears to be the outcome of historical causes, namely the different effect of the Pleistocene glaciations on the two oceans. And the latitudinal diversity gradient may be due to both kinds of causes acting together, but it is difficult if not impossible to assess their relative importance.

4. HORIZONTAL AND VERTICAL DISPERSAL IN THE SEA

As was remarked on page 138, two of the great contrasts between the marine and terrestrial biospheres are that the pelagic realm is three-dimensional and that it lacks a structured "vegetation." All inhabitants of the pelagic realm (as opposed to the benthic) can move; they can either swim or float. Currents can carry them immense distances. By rising or sinking they can move among environments that offer an array of different conditions of light and pressure and food-availability. No concrete, tangible barriers exist; in theory, all parts of the sea are accessible from all others.

Limits to the movement of individuals, or populations, are set by their tolerance limits for such variables as temperature and pressure. No doubt different groups differ greatly in the amplitudes of their tolerance ranges for various factors, and in the relative importance to them of different factors. For example, some jellyfish can live only in cold water but are indifferent to pressure. The three-dimensional shape of the geographic range of such a species thus forms a moderately thick, sagging sheet that dips to great depths below the warm surface waters of the tropics and emerges at the surface in high latitudes. In other words, it corresponds with the upper layers of the psychrosphere. Examples of animals with such ranges are members of the genera *Atolla* and *Periphylla,* both scyphozoans of the order Coronatae, the coronate mudusae (Lucas, 1974) (see page 278 and Figure 9.7).

The geographic ranges of many marine organisms are of great areal but small vertical extent. Inhabitants of different depth layers can be thought of as allopatrically distributed from the evolutionary point of view. Thus Marshall (1963) considers that, among fishes, speciation has resulted from competition within populations. He supposes that through evolutionary time, populations of many fish species occupying the shallow, productive waters over the continental shelf and upper continental slope increased to the point where pressure on living space forced some colonists into less productive waters. These excluded populations were forced first from shelf waters into the surface waters of the deep ocean, and then from the surface into successively deeper layers. As a result of this process, "the invaders would end by becoming isolated from the parent species" and evolutionary divergence would be possible. The excluded populations would become "part of the life of the great ocean gyres" and would no longer be able to exchange genes with their ancestral populations; (both quotations are from Marshall, 1963). Because of a shortage of food, dwarf mutants were probably selectively favored and this has led to the characteristically small size of deep-water fish species.

The new inhabitants of the deeper layers often became separated areally into populations completely isolated from each other by topographic barriers. For example, the Red and Mediterranean Seas are both deep basins separated by shallow sills from their neighboring oceans. The Red Sea has a maximum depth of 2800 m but the Strait of Bab-el-Mandeb joining it to the Indian Ocean is only 125 m deep. Likewise, the Mediterranean has a maximum depth of 4600 m, but the Strait of Gibraltar is nowhere deeper than 320 m. The shallow-water links via these two straits, between these basin seas and their respective oceans, probably date from the late Pliocene (Ekman, 1953). Populations forced into deep layers by competition will therefore have found themselves in isolated basins having their own local conditions. Gene exchange between deep-water populations of the Red Sea and Indian Ocean in the one case, and between the Mediterranean and the Atlantic in the other case, has thus been prevented.

The degree of divergence that has taken place has differed in the two cases, however. Physical conditions in the Red Sea differ markedly from those in the Indian Ocean; because of very high temperatures and the lack of inflowing rivers (the surrounding lands are all desert), salinity in the Red Sea is exceptionally high, at times over 40‰. The physical contrast between the Mediterranean and the Atlantic is not nearly so great. Correspondingly, the genetic divergence between "sister" populations of fishes of the Red Sea and Indian Ocean is probably much greater than that between sister populations of the Mediterranean and Atlantic. The Red Sea has a far larger proportion of endemic species than the Mediterranean (Briggs, 1974), but a smaller total number of species have been able to adapt to the more extreme Red Sea environment (Marshall, 1963).

The pelagic realm is the home not only of the nekton and the holoplankton, but also of the immature larval animals of the meroplankton, some of which settle down to a benthic life when they reach maturity. Dispersal over very long distances is therefore possible for them, at least in theory. They spend the larval stages of their lives in surface waters and hence their dispersals are not blocked by submarine sills and ridges, as are those of deep-water species.

It is possible, for example, for the veligers (advanced larvae) of some mollusk species to be transported across the Atlantic even though the adults are confined to the shelf benthos of the eastern and western shores. Thus although adult populations of such species are isolated from each other by the deep ocean, which they cannot cross, gene exchange is made possible by the easy trans-Atlantic movement of immature forms. Scheltema (1971, 1972) has examined over 800 plankton tows, taken from all over the tropical and warm-temperate North Atlantic, searching for the veligers of gastropods; his general conclusion is that transoceanic dispersal is a commonplace event. The frequency with which it happens must depend on the abundance of the offspring produced by shore-dwelling adults, on the probability that the veligers will be carried from coastal waters into the circulation of the open ocean, on their chances of surviving for the duration of the journey, and on their chances of shifting back from ocean to coastal waters when they reach the other side.

There are three trans-Atlantic surface currents at low latitudes. Two of these (the eastward-flowing North Atlantic Drift and the westward-flowing North Equatorial Current) are part of the North Atlantic gyre; the third is the westward-flowing South Equatorial Current. Scheltema found gastropod veligers in all these currents. Three species were common in all three currents; two (including *Tonna galea*; see Figure 5.10) were found all around the North Atlantic gyre, and more rarely in the South Equatorial current; and two were found in the North Atlantic Drift and the South Equatorial current.

As adults, these species have amphi-Atlantic distributions; they are members of the shelf benthos on both sides of the ocean. Scheltema (1972) found a direct relationship between the estimated frequency of trans-oceanic dispersal of the veligers and the degree of morphological similarity between populations on the two sides of the ocean. For example *Tonna galea* (family Tonnidae, tun shells) (see Figure 5.10) is found frequently in the plankton, and its eastern and western adult populations are indistinguishable. In contrast, *Thais haemastoma* (family Muricidae, dogwinkles), which probably drifts across the ocean much less frequently than *T. galea,* is represented on the two sides by morphologically distinguishable subspecies.

For a new species to evolve in the holoplankton, it is undoubtedly necessary for a single ancestral species-population to become separated into two or more isolated (or at least partly isolated) subpopulations which then diverge; this, presumably, is the mechanism by which nearly all speciation takes place. However, it is difficult to envisage how separation

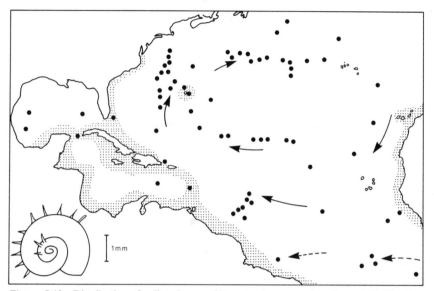

Figure 5.10 Distribution of veliger larvae of *Tonna galea*, a species of tun shell. The dots mark locations of plankton samples in which the species occurred. Postlarval stages are found in the areas shown stippled. Solid arrows, currents of the North Atlantic gyre; broken arrows, the South Equatorial current. Inset: larva of *T. galea*. (Redrawn from Scheltema, 1971.)

and isolation of portions of the holoplankton can happen; there are no barriers in the surface layers of the sea.

A possible separation process has been suggested by McGowan (1963). He supposes that as a plankton population is carried across the ocean surface by currents, from one marine climate into another, differential survival will ensure that the downstream population is selected *non*randomly from the upstream, source population. If we attach the labels A and B respectively to the source and destination areas and their populations, then it is seen that population B must diverge genetically from population A. The populations are prevented from mingling randomly because the current flows one way and thus their semi-isolation is maintained by an extrinsic cause. But although members of population B cannot return against the current to area A, selected members of population A, those that can survive the changed conditions, migrate continually into area B. McGowan considers that "eventual reproductive isolation could evolve," that is, that given time, populations A and B would become incapable of interbreeding even if the flow of the current did not keep them apart.

As an example of this process apparently in operation, McGowan

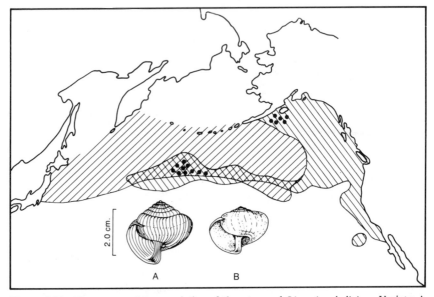

Figure 5.11 The ranges of two varieties of the pteropod *Limacina helicina*. Variety A occurs in the western and variety B in the eastern shaded areas. Dots show regions where AB intergrades are found. The current flows from west to east. (Redrawn from McGowan, 1963.)

(1963) describes the geographical ranges of two varieties of the species *Limacina helicina,* a pteropod (that is, one of the small planktic gastropod species known collectively as sea butterflies). Their ranges are shown in Figure 5.11. The "upstream" population, variety A, occurs in the western North Pacific and the "downstream" population, variety B, to the east of it. They differ morphologically in that variety A has high-spired, striated shells and variety B squat, smooth shells. There is strong evidence that the difference is genetic; intergrades are found in two regions. The temperature-salinity relations (TS curves, see page 141) of areas A and B are different but there is no sharp boundary; rather, conditions change gradually. The current flows from west to east. The curious feature of the geographic ranges is that variety B's range has a long tongue which extends westward from its main area *into* the current. The mechanism described above seems the only possible explanation for the existence of this tongue, which points in the upstream direction. *Limacina helicina* perhaps illustrates a mechanism that often leads to isolation and speciation even in the continuous, flowing surface waters of the sea.

Chapter Six

ISLAND BIOGEOGRAPHY

The biogeography of islands occupies a special place in the subject of biogeography as a whole. Whereas the continents and oceans of the world are few in number and each is unique, islands exist in teeming multitudes. Therefore they lend themselves to statistical study. The characteristic of an island's biota that is easiest to observe and quantify is the number of species it contains of some specified plant or animal group, for instance, the number of species of vascular plants, or of lizards, or of breeding birds, or of ants. Islands are numerous enough to permit the testing of theories intended to predict these numbers.

During the last two decades a large body of theory has grown up on this topic and it is discussed in Sections 1 and 2 of this chapter; the first section describes the theory in qualitative terms and the second deals with mathematical modeling of the theory in quantitative terms.

In its simplest form, the theory assumes that the numbers of species on islands result from processes too rapid to be affected by evolutionary change. Section 3 explores the effect on the theory of dropping this assumption and discusses the so-called taxon cycle, that is, the sequence of evolutionary stages through which a taxon is presumed to progress from its original appearance to its ultimate extinction. This evolutionary behavior is most clearly exhibited and most easily studied on island archipelagoes.

Another aspect of island biogeography, that of the colonization of oceanic islands by long-distance dispersal, is considered in Chapter 8, Section 3.

1. THE EQUILIBRIUM THEORY OF ISLAND BIOGEOGRAPHY

The Simple, Qualitative Version of the Theory

Consider a small island and its fauna of, for example, breeding birds. According to the equilibrium theory of biogeography (Preston, 1962; MacArthur and Wilson, 1963, 1967), the number of breeding bird species is repeatedly augmented by new immigrants from the nearest continent and repeatedly depleted by local extinctions. When the rates of immigration and extinction are equal, the bird fauna of the island is in a state of dynamic equilibrium. That is, the number of species on the island achieves a constant equilibrium level, $\hat{S}$, and remains there, apart from temporary deviations due to chance. However, although the number of species remains (approximately) constant, there is a continual turnover of species. During an interval of time, the composition of the fauna will change, owing to the loss, by local extinction, of some species and the gain, by immigration, of an equal number of others. The equilibrium turnover rate (the replacement rate) may be large or small; it is equal to the equilibrium immigration and extinction rates, which are equal to each other.

In its simplest form, the theory makes only four further assumptions. These are that:

1. The immigration rate, I, decreases as the number of species, S, already present on the island increases.

2. For given S, I depends on the distance of the island from the mainland; the nearer the island lies to the mainland from which the immigrants come, the greater the magnitude of I.

3. The extinction rate, E, increases as S increases.

4. For given S, E depends on the area of the island; the smaller the island, and hence the smaller the population of each species that it can support, the greater the risk of any one of them becoming extinct.

The theory can be portrayed diagrammatically as in Figure 6.1. The two curves in the graph show how I and E vary with S for some one particular island of given area and given distance from the mainland. When the island's fauna is at equilibrium, $\hat{S}$ species will be present. If, by chance, immigrations outnumber extinctions for a short period, causing S to increase to a value greater than $\hat{S}$, I will tend to decrease and E increase, forcing S back toward its equilibrium level; correspondingly, a chance fall in S will have the opposite effect, and will again be followed by restoration of the equilibrium.

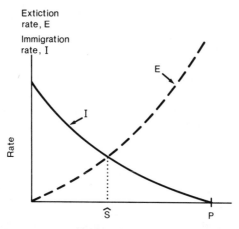

Number of species, S

Figure 6.1   Model showing how the number of species on an island results from a dynamic equilibrium between extinctions (E) and immigrations (I) which occur at rates varying with S, the number of species on the island at any time. When $I = E, S = \hat{S}$, the equilibrium value. The number of species in the mainland pool is P. (Redrawn from MacArthur and Wilson, 1963.)

The simplest version of the theory assumes that $\hat{S}$ is wholly determined by three factors: the size, say P, of the mainland species-pool, that is, the number of species in the source region from which the immigrant species come; the area of the island; and its distance from the mainland. Consider again the island whose immigration and extinction rates are shown in Figure 6.1. A larger island at the same distance from the mainland would have a lower extinction curve (dashed line) but the same immigration curve (solid line); therefore the point of intersection of the curves would be shifted to the right, or equivalently, $\hat{S}$ would be greater. An island of the same size but farther from the mainland would have the same extinction curve but a lower immigration curve (see Figure 6.3a); therefore the point of intersection of the curves would be shifted to the left, or equivalently, $\hat{S}$ would be lower.

The foregoing paragraphs summarize the theory in its original simple form. We now consider objections to it, and elaborations and refinements that might be made to improve it.

Equilibrium Theory: Complications and Refinements

The attractiveness of the theory as a simple, general, testable natural law is obvious. It has inspired a large volume of research since it was first proposed (Preston, 1962; MacArthur and Wilson, 1963, 1967). Indeed, its appearance triggered a tremendous and continuing surge of interest in biogeography; in the words of Simberloff (1974), who reviewed developments in the theory during its first decade, it has revolutionized the subject.

Reaction to the theory has ranged from uncritical acceptance to flat rejection. Whether to accept, modify, or reject the theory depends on whether one regards its assumptions as acceptable, as too simple but capable of improvement, or as oversimplified to the point of absurdity. The numbered paragraphs below deal with eight chief objections to the simple theory; some of them could be overcome by appropriate refinements to the theory. (It will be seen that some of the points mentioned overlap each other.)

1. The opinion that the theory is so oversimplified as to be valueless is held by, among others, Sauer (1969) and Lack (1970). According to Sauer, for example, the model "filters out the interpretable signal instead of the random noise."

2. The model assumes that new species appear on an island only as a result of immigration and not from autochthonous evolution. Equivalently it assumes that evolution rates are so slow relative to migration rates that they can safely be ignored. This is probably untrue. The point is discussed in more detail in Section 3.

3. The simple model assumes that I decreases monotonically as S increases. Although this may be true for many groups of animal species, it is probably incorrect for plants (bryophytes and tracheophytes) treated as a single group. Imagine a bare, lifeless island, for example one whose vegetation has been completely destroyed by a volcanic eruption. It is intuitively reasonable to suppose that the first plant colonizers, those able to establish themselves in dry mineral soil, will immigrate at a low rate, but once these pioneers begin their soil-forming function, new species of colonists will start to arrive much more rapidly. Thus I would increase with increasing S for low values of S as shown in Figure 6.2.

4. A simplification that can easily be allowed for if necessary is the assumption that I depends only on an island's distance from the mainland. It seems reasonable to postulate that a large island is more likely than a small one to lie in the path of such migrating animals as birds and insects. Of course, it does not necessarily follow that the effect is appreciable and that incorporating it in the model will yield any perceptible improvement (see Section 2).

5. The assumption that E depends only on an island's area may also be faulty. Immigration and extinction are not necessarily independent; if a

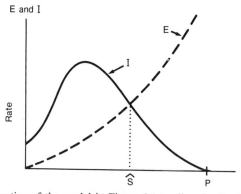

Figure 6.2 Modification of the model in Figure 6.1 to allow for the fact that with some organisms (for example, plants), I is likely to increase with S when S is low.

species-population is dwindling, extinction may be temporarily staved off by an influx of immigrants, which will enlarge the population demographically and enrich it genetically. Such an influx is more likely into an island near the mainland than into one far away. Brown and Kodric-Brown (1977) have called this the "rescue effect." It is shown diagrammatically in Figure 6.3. As may be seen, if there is no rescue effect, the species-turnover rate is greater on a near island than on a far island of equal area. But if distance is assumed to affect extinction as well as immigration, then the turnover rate can be greater on a far island as often appears to be true in nature.

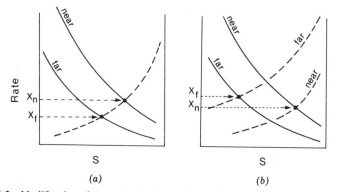

Figure 6.3 Modification of the model in Figure 6.1 to allow for the fact that E is greater on far than on near islands because of a "rescue" effect. The curves relate to islands of the same area. X_n and X_f are the species-turnover rates on a near and a far island respectively. (a) Simple model with E independent of distance, $X_n > X_f$. (b) Modified model with E dependent on distance, $X_f > X_n$. (Adapted from Brown and Kodric-Brown, 1977.)

6. The assumption that an island's area determines the number of species it can support may be an oversimplification. Mountainous islands presumably provide a greater range of different habitats than do low-lying islands. Thus topography as well as area probably has a marked effect on $\hat{S}$.

7. The theory treats all species as equal. It assumes that the probability of immigrating successfully, or of going extinct, is the same for all species in the group under consideration. Whether this assumption is an acceptable approximation to the truth depends, therefore, on the group concerned. It may hold well enough for restrictively defined groups, but for large groups it probably fails. As an example, suppose the group under consideration were all seed plants. Then, as Whitehead and Jones (1969) argue, it would be incorrect to treat a single immigration curve as applicable to both strand and nonstrand plants (see Figure 6.4). These two classes of plants differ in two important respects; (below, the relevant symbols are subscripted with S and N, for strand and nonstrand). First, strand species tend to immigrate faster than nonstrand species and at a rate almost independent of distance from the mainland source, because many species are dispersed by ocean currents. Second, the number of strand species in the mainland pool, P_S, is often low and $\hat{S}_S = P_S$ approximately for quite small islands. By contrast, $\hat{S}_N$ is zero for extremely small islands since $\hat{S}_N > 0$ requires that an island be large enough (with area $> A_{min}$, say) for a permanent lens of fresh ground water to form; without one, nonstrand species cannot persist. The foregoing arguments show, of course, only that the simple theory is inapplicable to heterogeneous groups of species. They do not show whether or not any groups exist that are large enough to be interesting and homogeneous enough for the theory to apply.

8. The simple model treats $\hat{S}$ as constant for a particular island and a particular species group. If the observed number of species, S_{ob}, and the observed turnover rate on an island, remain constant over a long period, one is led to conclude that the model applies and that the island's species complement (of the group concerned) is in a state of dynamic equilibrium. However, the model is not disproved if S_{ob} is found to vary. Indeed, two interpretations can be put on such variation. Either the island's $\hat{S}$ is a constant toward which S_{ob} is tending but the biota is in a nonequilibrium state owing to a recent disturbance, or the island's biota *is* at equilibrium but $\hat{S}$ itself varies with time because of secular variation in climate (see Figure 6.5). Often, of course, both interpretations may apply. If the approach of S_{ob} to $\hat{S}$ is slow relative to the rate at which $\hat{S}$ varies, $| S_{ob} - \hat{S} |$ will vary unpredictably.

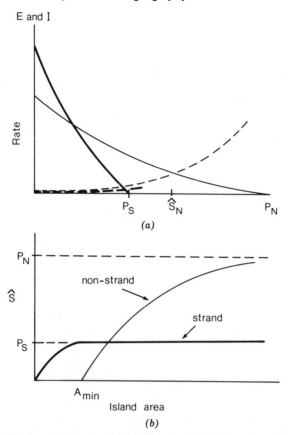

Figure 6.4 (*a*) The immigration curves (solid) and extinction curves (dashed) for strand plants (heavy lines) and nonstrand plants (fine lines). For the strand plants, $\hat{S}_S = P_S$ approximately. (*b*) The relation of $\hat{S}$ to island area for strand and nonstrand plants.

Evidently, therefore, the model suffers from the defect that it is difficult to disprove. In critical vein, one might say that if the assumptions hold, the conclusions follow inevitably; hence that the model *must* apply except to the exceptions. This criticism could be aimed at all ecological models, of course, not merely the equilibrium model of island biogeography we are here considering. Rather than debating the propriety of model-building, however, it is more useful to consider actual biogeographical observations that have been inspired by the island model and the insights they have led to.

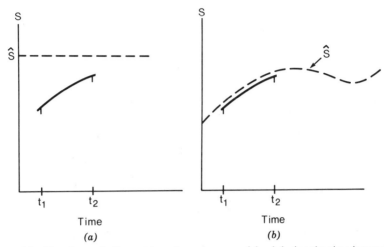

Figure 6.5 The change in the number of species on an island during the time interval (t_1, t_2). (a) The island's biota is not in equilibrium and S is approaching $\hat{S}$. (b) The island's biota is in equilibrium, but $\hat{S}$ varies with time. (Note that S_{ob} can also exhibit a mixture of the two behaviors.)

Equilibrium Theory: Tests and Interpretations

The equilibrium model is most likely to be applicable to fairly narrowly defined groups of species on sets of fairly similar small islands. Where it applies, one expects to find that the number of species (of the taxonomic group concerned) on an island (of the island group concerned) is largely determined by the island's area and by its degree of isolation from the mainland. The effects of other factors are expected to be comparatively minor, though not necessarily imperceptible.

Tests of the theory—or, rather, of its applicability in particular contexts—have been numerous. The breeding bird species of Pacific island groups and archipelagoes have been especially well studied (see Simberloff, 1974, for references). In many cases very strong evidence has been obtained suggesting that species numbers are indeed governed by the two factors, area and isolation, that the theory stresses. A more detailed, mathematical, account of the correspondence between theory and observation in one particular case is given in Section 2.

Birds have constituted the test animals in a great many investigations because they are conspicuous and well known, and because species lists of the birds on many islands at many different times can be found in the literature. The islands off the coast of California have been particularly well studied and their bird faunas have been observed and recorded over many decades.

Conclusions based on these data are controversial, however. The theory is worth applying only to resident species, that is, to species that are adapted members of the island's biotic communities. In the case of birds, this means breeding birds. If transients, migrants, and miscellaneous strays are counted as belonging to an island's bird fauna, one may indeed observe a balance between arrivals and departures but, as Lynch and Johnson (1974) remark, this merely amounts to an observation of the "unsurprising fact that transient species are, in fact, transient." To test the applicability of the model to birds, therefore, a species should be counted as an immigrant only if it remains in an island to breed for at least one reproductive cycle; and as having gone extinct only if it disappears for at least one reproductive cycle; (it is possible for a species to immigrate and become extinct repeatedly, of course). According to Lynch and Johnson these rules have not been observed in studies of the birds on islands off the California coast; as a result, they believe, species turnover rates at equilibrium have been grossly overestimated and this has led to an unwarranted stress on the dynamic character of the species equilibrium.

With plants, also, a distinction should be made between true colonizers and transients, as was done by Heatwole and Levins (1973). They investigated the plant species equilibrium of a small sand cay off Puerto Rico and obtained values for $\hat{S}$ of 11.2 and 6.8 depending on whether all species were tallied, or established (reproducing) species only; (these values are the means of several estimates). The corresponding turnover rates were approximately 16 and 1.2 species per year.

A very careful and detailed test of the theory was carried out by Simberloff and Wilson (see Simberloff, 1969, and references therein). In this work, the islands consisted of mangrove clumps rooted in the mud below shallow, tidal water in Florida Bay; their areas ranged from 75 to 500 m^2 and their distances from the mainland from 2 to 1200 m. The species considered were the terrestrial invertebrates (insects, arachnids, chilopods, diplopods, and the like) inhabiting the plants; the islands lacked dry land. To test the theory, several islands were completely defaunated, by fumigation, and the subsequent natural recolonization was then closely monitored.

The results conformed very closely with theoretical predictions. Good correlation between $\hat{S}$ and island area was found, and this raises and perhaps answers the following question: to what extent is the dependence of $\hat{S}$ on area merely the outcome of the fact that larger islands usually have a larger variety of habitats? In the case of bird species on fairly large islands, especially hilly or mountainous islands, it seems certain that part of the correlation between $\hat{S}$ and area arises from a correlation between habitat diversity and area. However, Simberloff and Wilson's work clearly demonstrates that area per se controls $\hat{S}$ when, as in the case of the mangrove islands, all possible microhabitats are found on every island no matter how small.

Another point to emerge from the mangrove islands experiment is that there is more than one kind of species equilibrium, even when ecological (as opposed to evolutionary) equilibria are considered. The first equilibrium to be attained by the animals as they recolonized the defaunated islands was a temporary *noninteractive* equilibrium: immigrations and extinctions reached a balance while populations were still too small for interspecific competition to have noticeable effects. Subsequently, the number of species fell slightly, to a presumably permanent *interactive* equilibrium number: populations had now grown to the point at which competitive exclusion had caused the local extinction of some of the original colonizers.

When an island fauna had reached interactive equilibrium, it was assumed that the island was saturated with species and that the number would remain approximately constant. However, while this equilibrium number was being approached, and perhaps after it had been attained, adjustments to the taxonomic structure of the island community (its species composition) were also in progress, leading to an *assortative* equilibrium (Simberloff, 1974). That is, qualitative changes took place in each island's species complement: extinctions tended to be more frequent among ill-adapted species, and successful immigrations more frequent among well-adapted ones. This sorting process finally led to the establishment on each island of a coadapted set of species, which was not a random subset of the species in the mainland pool.

The observation that the final equilibrium constellation of species on each island was a coadapted set rather than a random assortment of the species available is due to Heatwole and Levins (1972, 1973). They showed that the animal communities on all the islands, both before experimental defaunation and after complete recovery, were characterized by a constant trophic structure. This was established and maintained in spite of great variation from island to island in the taxonomic structures (species compositions) of their separate communities. The species were assignable to seven trophic categories. The categories, and the proportions of species in them on each island, were as follows: herbivores, 0.36; predators, 0.22; ants (whose trophic status is uncertain), 0.18; parasites, 0.07; detritus feeders, 0.07; wood borers, 0.05; scavengers, 0.03; and undetermined, 0.02. This trophic structure tended to be the same for all islands, and to be temporally stable. It persisted despite the instability of the communities' taxonomic structures, and can be regarded as a biogeographic characteristic of the mangrove islands of the region (the Florida coast).

Probably it is true in general that islands with particular geographical and ecological properties are characterized by faunas with particular trophic structures. Another example described by Heatwole and Levins (1973; and see Heatwole, 1971) concerns the animals found on bare, unvegetated cays in the Coral Sea. Five trophic categories were recognized and the proportions of species in them were as follows: detritus feeders, 0.47; ectoparasites of sea birds, separated from their hosts, 0.27; scavengers, 0.16; predators 0.07; and herbivores, 0.02. Of course a particular trophic structure can be described as characteristic of a whole island only if the island is small and ecologically uniform.

In any case, the fauna of a small island can have a trophic structure determined by ecological factors, while the equilibrium number of species is determined by the area and isolation of the island. The two concepts are not inconsistent.

Far more observational work on island biogeography has been done in the tropics and subtropics than in higher latitudes. Abbott and Grant (1976) speculate that, because of irregularly fluctuating climatic conditions, islands in high latitudes are unlikely to have temporally constant values of $\hat{S}$. In such conditions, an island biota may either never achieve equilibrium, or else may be in a state of "wandering equilibrium" as in Figure 6.5b.

Abbott (1974) also found that the relation between species number and island area was much less close for islands in high latitudes than in low. He investigated the numbers of species of plants, of insects, and of birds on 19 remote oceanic islands whose latitudes ranged from 37°S to 61°S, and judged the dependence of species numbers on several predictor variables, among them area, elevation, temperature of the coldest month, and degree of isolation. Island area turned out to be a comparatively unimportant predictor of species number in all three biotic groups. For plant species, the single most important factor was temperature; for insects and also for birds, it was the number of plant species. The number of bird species was not appreciably affected by the number of insect species even though the majority of the birds were passerine insectivores. The greater importance of plants than of insects in controlling the number of bird species is probably because plants create the habitat for both insects and birds; the factor of overriding importance for both groups of animals may be the number of different habitats available. Indeed, it is conceivable that the number of plant species on an island, and hence the number of habitats, is determined by temperature in cold parts of the world and by island area in warm. But the data are still insufficient for such generalizations to be more than guesswork at present.

*2. THE QUANTITATIVE THEORY OF ISLAND BIOGEOGRAPHY

Thus far we have considered the equilibrium theory of biogeography only in qualitative terms. To make it quantitative, it is necessary to express the immigration and extinction rates at any instant as functions of the variables S, A, and D which denote, respectively, the number of species present at that instant, the area of the island, and its distance from the mainland source of immigrants. Values of $\hat{S}$ for given values of A and D can then be predicted and the predictions tested. It also becomes possible to predict the time required for S to recover after being temporarily in a state of disequilibrium. We discuss the latter topic first.

The Relaxation Time for an Island in Disequilibrium

An island becomes abruptly undersaturated with species ($S < \hat{S}$) if part or all of its biota is destroyed by some cataclysm such as a cyclone or a volcanic eruption (or because of experimental defaunation); and it becomes oversaturated ($S > \hat{S}$) if its area is suddenly reduced, for example, by a rise in sea level. In either case, the biota is thrown into disequilibrium. This implies that immediately after the perturbation, the immigration rate, I, and the extinction rate, E, are unequal. However, the imbalance between I and E leads to the restoration of equilibrium. To predict the length of time that recovery will take, it is necessary to express E and I as functions of $S(t)$, the number of species on the island at time t (we now write $S(t)$ rather than S to emphasize its dependence on time).

The simplest possible assumptions are that E is proportional to $S(t)$ and that I is proportional to the amount by which $S(t)$ falls short of P, the number of species in the mainland pool. That is, for a given island,

$$E = \epsilon S(t) \quad \text{and} \quad I = \iota(P - S(t))$$

where ϵ and ι are constants of proportionality. The assumptions amount to replacing the curves in Figure 6.1 with straight lines.

The rate of change of $S(t)$ is the difference between I and E. That is,

$$\frac{dS(t)}{dt} = \iota P - (\iota + \epsilon)S(t).$$

Therefore

$$\frac{dS(t)}{\iota P - (\iota + \epsilon)S(t)} = dt.$$

Integrating gives

$$\frac{-1}{\iota + \epsilon} \ln[\iota P - (\iota + \epsilon)S(t)] = t + \ln K$$

where the constant of integration has been defined as $\ln K$.
    Writing $\iota + \epsilon = \alpha$ (for conciseness) yields

$$\ln (\iota P - \alpha S(t)) = -\alpha t + \ln K.$$

(The symbol for the constant has been left unchanged.)
Thus

$$S(t) = \frac{1}{\alpha} (\iota P - K e^{-\alpha t}). \tag{6.1}$$

Now consider $S(0)$, the number of species on the island immediately
following the initial perturbation at $t = 0$. Clearly,

$$S(0) = \frac{\iota P}{\alpha} - \frac{K}{\alpha} \qquad \text{whence } K = \iota P - \alpha S(0).$$

Next note that as $t \to \infty$, $S(t) \to \hat{S}$ and hence that

$$\hat{S} = \frac{\iota P}{\alpha}.$$

Then, from (6.1),

$$S(t) = \frac{\iota P}{\alpha} - \frac{K}{\alpha} e^{-\alpha t} = \frac{\iota P}{\alpha} - \left(\frac{\iota P}{\alpha} - S(0)\right) e^{-\alpha t}$$

$$= \hat{S} - (\hat{S} - S(0))e^{-\alpha t}.$$

It follows that

$$\frac{\hat{S} - S(t)}{\hat{S} - S(0)} = e^{-\alpha t}. \tag{6.2}$$

Notice that the lefthand side of (6.2) is the deviation of $S(t)$ from equilibrium at time t as a fraction of its initial deviation due to the perturbation at $t = 0$. The perturbation may have caused the island to be undersaturated initially, in which case $S(0) < S(t) < \hat{S}$ for all t in $(0, \infty)$; or, if the perturbation caused oversaturation initially, $S(0) > S(t) > \hat{S}$ for all t in $(0, \infty)$.

We now inquire how long it will take for an island in disequilibrium to make some chosen fraction, say 90%, of a complete recovery. (Since the approach to equilibrium is asymptotic, the time to full recovery is theoretically infinite.) The answer, denoted by $t_{0.90}$ (MacArthur and Wilson, 1967), depends only on the constants ϵ and ι. Equivalently $t_{0.90}$ can be defined as the time required for a deviation from equilibrium to decrease to 10% of its initial magnitude.

Putting the lefthand side of (6.2) equal to 0.10, and recalling that $\alpha = \epsilon + \iota$, it is seen that

$$0.10 = \exp[-(\epsilon + \iota)t_{0.90}],$$

or

$$t_{0.90} = \frac{-\ln 0.10}{\epsilon + \iota} = \frac{2.3026}{\epsilon + \iota}.$$

Diamond (1972) defined the *relaxation time*, t_r, of an island biota as the time required for a deviation from equilibrium to decrease to $e^{-1} = 36.8\%$ of its initial magnitude. That is, he put

$$\exp[-1] = \exp[-(\epsilon + \iota)t_r],$$

whence

$$t_r = \frac{1}{\epsilon + \iota} = \frac{t_{0.90}}{2.3026}.$$

Notice that 95% recovery takes place in three relaxation times.

It must be emphasized that the foregoing discussion applies *only if the assumptions hold*. The available empirical evidence suggests that, in fact, ι and ϵ are not constants. It appears that ι decreases and ϵ increases with increasing S.

These inferences are intuitively reasonable. Thus, given an empty island, new species are likely to accumulate rapidly at first, as easily dispersed organisms pour in from the mainland pool; the later arrivals will tend to be the poor dispersers, and the immigration of new species will therefore slow down for this reason as well as because the number of possible new immigrants is steadily dwindling. The inferred increase of ϵ with increasing S can probably be attributed to the increasing likelihood that a species will suffer competitive exclusion as the number of possible competitors grows.

In the models to be discussed next, therefore, the assumptions that $I = \iota S(t)$ and $E = \epsilon(P - S(t))$ are discarded as too simple.

Gilpin and Diamond's Models

Gilpin and Diamond (1976) devised, and compared with real data, no fewer than 13 different versions of the equilibrium model, that is, 13 pairs of functional forms for the rates E and I. The models were devised to fit, and were tested against, data on the lowland bird fauna of 52 islands in the Solomon Islands archipelago just east of New Guinea. In all their models the authors assumed that E was a function of S and A, say $E(S, A)$, that behaves as a product of two functions, one of S alone, and the other of A alone. They therefore wrote

$$E(S, A) = g(S) e(A).$$

In 10 of their models they made a similar assumption about I, which they regarded as depending on S and D, and wrote

$$I(S, D) = h(S) i(D).$$

[In their remaining three models, they permit I to depend on A as well as on S and D and put $I(S, A, D) = h(S) i(A, D)$.]

They then sought for intuitively reasonable expressions for the functions $g(S)$, $e(A)$, $h(S)$, and $i(D)$. It was also desired that these functions should embody as few unknown parameters as possible. The numbers of parameters in the various models ranged from two to seven.

We here discuss their Model 5 (with three parameters) and then, briefly, their Model 10 (with six).

Consider $g(S)$ first. As argued above, and as implicitly assumed by representing the relation between E and S by a concave curve (the dashed curve) in Figure 6.1, it seems reasonable to assume that $g(S)$ increases with S at a rate that itself increases with S. That is, we require that the first and second derivatives of $g(S)$, namely $g'(S)$ and $g''(S)$, should obey the relations $g'(S) > 0$, $g''(S) > 0$. Hence we try

$$g(S) = E_0 s^n$$

with $n > 1$; E_0 is a constant of proportionality.

Next, suppose that $E(S, A)$ is inversely proportional to A and put $e(A) = 1/A$; there is no need to introduce another constant of proportionality since $e(A)$ appears only as a factor multiplying $g(S)$. Then

$$E(S, A) = \frac{E_0 s^n}{A} \qquad (6.3)$$

is a two-parameter model for the extinction curve, with parameters E_0 and n.

Now consider $h(S)$ and $i(D)$. For bird species it is reasonable to assume that the immigration curve relating I and S is concave as shown in Figure 6.1; (this assumption would not be reasonable for plant species; see Figure 6.2). That is, we assume that $h'(S) < 0$ and $h''(S) > 0$. Further, $h(S)$ is presumably a decreasing function of S/P, the fraction of the species from the mainland pool that has already reached the island. Hence we try

$$h(S) = I_0 (1 - S/P)^m$$

with $m > 1$, and I_0 as a constant of proportionality.

Lastly we come to $i(D)$. The expression that Gilpin and Diamond found most economical of parameters while at the same time giving an acceptable fit to the data is

$$i(D) = \exp\left[-\frac{\sqrt{D}}{D_0}\right]$$

with the single parameter D_0.
Then

$$I(S, D) = I_0\left(1 - \frac{S}{P}\right)^m \exp\left[-\frac{\sqrt{D}}{D_0}\right]. \qquad (6.4)$$

Equations (6.3) and (6.4) contain five parameters, namely E_0, I_0, n, m, and D_0. The number of species in the mainland pool, P, was independently known and thus did not need to be inferred from the data. To solve the equations for $\hat{S}$, which is the purpose of the modeling, entails putting $E(S, A) = I(S, D)$, or equivalently

$$\frac{E_0 S^n}{A} = I_0 \left(1 - \frac{S}{P}\right)^m \exp\left[-\frac{\sqrt{D}}{D_0}\right].$$

It is seen that there is no point in keeping I_0 and E_0 separate. Therefore we may put $I_0 = 1$ and retain as an unknown parameter only E_0, which Gilpin and Diamond now replace with R.

Further empirical testing showed that a model with $m = 2n$ worked well. This accorded with the fact, supported by independent evidence, that immigration curves tended to be far more concave than extinction curves.

The three-parameter model (Model 5) therefore becomes

$$E(S, A) = \frac{RS^n}{A} \qquad (6.3')$$

and

$$I(S, D) = \left(1 - \frac{S}{P}\right)^{2n} \exp\left[-\frac{\sqrt{D}}{D_0}\right] \qquad (6.4')$$

with parameters R, n, and D_0. For the lowland birds of the Solomon Islands, it was known that $P = 106$.

The parameter values that gave the best fit of the model to these data were $R = 1.49 \times 10^{-5}$, $n = 2.37$, $D_0 = 2.11$. Figure 6.6 shows a representative pair of immigration and extinction curves. They relate to an island of area $A = 10$ mi^2 at a distance $D = 25$ mi from the mainland (New Guinea), and were obtained by substituting these values of A and D, and the fitted values of R, n, and D_0, into (6.3') and (6.4'). As already remarked, it was known beforehand that $P = 106$.

The model fitted the data well. It explained 96.8% of the observed variance in S. A six-parameter model (Gilpin and Diamond's Model 10) gave a slightly, but significantly, better fit and accounted for 97.9% of the observed variance. This more complicated model differed from Model 5 in three respects: (1) It did not assume that m [in (6.4)] could be approximated by $2n$; instead, m was determined separately. (2) In the expression for $i(D)$, D was replaced by D^y with $y \neq 0.5$; again, y was determined separately, to maximize the fit. (3) It was assumed that A affected I as well as E. The effect was allowed for by putting $i(D, A) = \exp[-D^y/D_0 A^v]$ in place of $i(D)$ and determining v so as to maximize the fit.

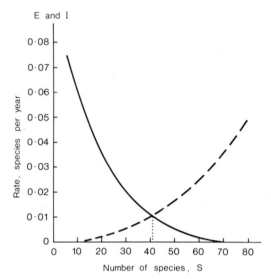

Figure 6.6 The immigration curve (solid) and extinction curve (dashed) of Gilpin and Diamond's (1976) three-parameter model with parameters $R = 1.49 \times 10^{-5}$, $n = 2.37$, $D_0 = 2.11$. The size of the mainland pool of species is known to be $P = 106$. The area of the island here considered is $A = 10$ mi^2 and its distance from the mainland is $D = 25$ mi. The curves intersect at $\hat{S} = 40.4$.

The excellent fit of the models to the data implies that almost all variation in S is due, in this body of data, to variation in A and D. It should be observed that, in choosing data against which to test their models, Gilpin and Diamond deliberately selected a group of islands, and a group of organisms (lowland birds), for which it was reasonable to postulate that S was indeed almost wholly determined by A and D. The islands all have a similar history, and the choice of lowland birds as the faunal group to be considered ensured that altitude should not be one of the factors affecting S.

3. LONG-TERM CHANGES IN ISLAND BIOTAS

In discussing island biogeography, it has been assumed up to this point that immigration and extinction rates are so high that events and conditions more than a few centuries in the past have left no trace on present-day island biotas; further, that all the species on an island are immigrants from elsewhere and that none have evolved in situ. Both these assumptions are unreasonable, and we now explore island biogeography from the historical point of view.

First it is necessary to emphasize the biogeographic distinction between continental and oceanic islands. A continental island is no more than a detached fragment of a continent, from which it has become separated by a rise in sea level (eustatic or isostatic) at some time in the past. Its biota, likewise, is merely a detached fragment of the biota of the parent continent. An oceanic island, in contrast, rises out of the deep ocean, usually far from land, and has never been linked to a continent by a land bridge. Its original biota, therefore, can have reached it only by transoceanic dispersal. Thus all its species are either long-distance dispersers (this includes species brought in by human agency) or the autochthonously evolved descendants of long-distance dispersers.

Continental Islands

When it first becomes separated from the mainland, a continental island has an oversaturated biota, according to equilibrium theory. Local extinctions will exceed immigrations for a period, until the number of species in the island falls to the equilibrium level appropriate to the island's area and distance from the mainland; this distance is the width of the flooded channel that has cut it off from the mainland.

Very clear evidence of this loss of species by oversaturated islands has been given by Wilcox (1978). He considered the lizard faunas of 17 islands off the coast of Baja California. The islands were known to have been part of the North American continent during the Pleistocene, and have been turned into islands by the eustatic rise in sea level caused by the melting of the Wisconsin ice sheets. Wilcox inferred the ages of the islands from a knowledge of the depths of the channels separating them from the mainland and the rate at which sea level has risen. The ages range from 5800 to 12,000 years. He found that the regression model of best fit ($r^2 = 0.91$) expressed S, the number of species on an island, as a function of the island's age, area, and latitude (which presumably affected climate and hence vegetation and its concomitant, habitat diversity). The relation between the residual variation in S after allowing for the regression on area and latitude thus shows the dependence of S on island age; it is shown in Figure 6.7. The way in which S falls with increasing age amounts to most persuasive evidence that the islands' lizard faunas are in the process of dwindling toward the equilibrium levels appropriate to their relatively recent island status.

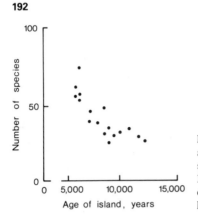

Figure 6.7 The relation between time since separation from the mainland and number of lizard species for 17 continental islands off the coast of Baja California. The numbers of species have been corrected for variation due to the islands' areas and latitudes. (Redrawn from Wilcox, 1978.)

Oceanic Islands and Endemism

Now consider oceanic islands. Their biotas have never been physically part of a larger, continental biota and must consist of immigrants and the descendants of immigrants. Some of the latter will have differentiated to yield new species or even genera, endemic to an individual island or to a group of islands.

The likelihood that a species-population in an isolated island will undergo fragmentation and speciation depends on the characteristics of the species concerned. Autochthonously evolved endemics are of such great importance in biogeography, and their evolutionary histories and geographic ranges are so carefully studied, that it is often forgotten that a great many island populations are conspecific with mainland populations. Probably these are comparatively recent immigrants and may have replaced an earlier population of the same species that chanced to go extinct.

A small species-population, confined to an island, can easily die out. There is the ever-present risk that deaths will chance to exceed births until no survivors remain. Also, small populations have small gene pools; the risk is therefore great that a change in environment will occur to which no population members are adapted. Further, homozygosity is likely to be high and is itself selectively disadvantageous (Mayr, 1970). However, such extinction is only local and may therefore be temporary. Reinforcements of the same species may again invade from the mainland and found a replacement population. A constant turnover of populations of one species would explain, to quote Darlington (1957), "the apparently rather recent age of most island faunas and the fact that vertebrates occur on islands in proportion to probable power of crossing salt water, not in proportion to geologic age." This may explain why autochthonously evolved endemics are uncommon in some taxonomic groups and on some islands.

Nevertheless, endemic species are very common among certain groups. For example, more than 20% of the land bird species on the following islands are endemic (Mayr, 1965): Madagascar, Jamaica, Hispaniola, Cuba, New Caledonia, Kauai (one of the Hawaiian Islands), Principe, and Sao Thome (the latter two are in the Gulf of Guinea). On some of the smaller islands of the Galapagos archipelago, 100% of the plant species are endemic to the archipelago (Johnson and Raven, 1973). Comparisons between disparate groups may be misleading, however, since so much depends on the taxonomic treatment of the groups considered.

Within groups that (it is hoped) have had consistent taxonomic treatment, differing degrees of endemism can be recognized. Thus Good (1974) notes that although there are some endemic species of plants in the Galapagos Archipelago, they are closely related to species in mainland South America and there are very few endemic genera. A far higher level of endemism is shown by the flora of Juan Fernandez, even though it is no farther from the coast of Chile than the Galapagos Islands are from the coast of Ecuador; it has numerous endemic genera and even one endemic family. The contrast is such that Good recognizes Juan Fernandez as a floristic region in its own right, one of the 37 into which he classifies the world's flora; but the Galapagos Islands are included in the Andean region.

The number of endemic species of a particular group on an island must obviously depend on many factors, especially the age of the island and its distance from a mainland source of immigrants. It also depends on the island's area. One could, indeed, entertain the hypothesis that the number of endemics on an island tends to a dynamic equilibrium in the same way that the total number of species (of which the endemics are a subset) is believed to tend to a dynamic equilibrium. Examples of the relation between percentage endemism and island area were given by Mayr (1965) in a study of the land birds on 29 oceanic islands scattered all over the world. He grouped the islands into four sets, depending on their locations relative to other islands and to the nearest continent, and found that the percentages of endemic species increased as island area increased in all four sets. The result for 14 islands in scattered archipelagos is shown in Figure 6.8a. Mayr ascribes the relation between endemism and area to the fact that the turnover rate for species-populations is presumably much greater on small islands than on large. Populations on small islands tend to go extinct before they have had time to diverge appreciably from their mainland ancestors. The larger an island, the greater the probability that a

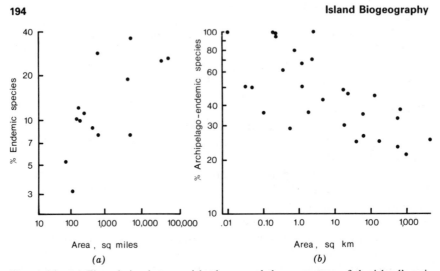

Figure 6.8 (*a*) The relation between island area and the percentage of the island's avifauna *endemic to that island* for 14 islands belonging to scattered archipelagos. (Redrawn from Mayr, 1965.) (*b*) The relation, within a compact archipelago, between island area and the percentage of its flora consisting of species *endemic to the archipelago,* for 29 islands in the Galapagos Archipelago. (Data from Johnson and Raven, 1973.)

resident population will persist long enough to develop into a distinct, endemic species. The relation between percentage endemism and area thus holds for low-latitude islands only, not for islands in colder climates which experienced complete faunal turnovers during the Pleistocene glaciations.

It is interesting to consider also the number of species that are endemic to an island archipelago, and the way in which such species are distributed among the archipelago's member islands. Johnson and Raven (1973) have made such a study, on plant species endemic to the Galapagos Archipelago. They found that 26% of the total flora are Galapagos endemics, that is, are endemic to the archipelago as a whole. They also found that the percentage of Galapagos endemics in the individual island floras decreases with increasing island area (see Figure 6.8*b*). This finding is unexpected and is probably not true of archipelagos in general. Johnson and Raven attribute the result to the climatic history of the Galapagos region, and hence to the past availability of different habitats. In tropical islands, the largest proportion of endemics is usually found among the species of moist upland areas; they are species that have evolved from related forms occurring at lower and usually drier elevations. This is not the case in the Galapagos Islands, however, because the uplands have not been moist during a long enough period for many endemics to evolve. The islands were comparatively arid at the end of the Pleistocene; moist uplands such as are now found probably came into existence no earlier than 10,000 years ago and have been colonized by plants from the mainland that have not yet had time to diverge. The habitat in which most of the Galapagos endemics occur is the arid lowland immediately inland of the strand zone which surrounds each island. This arid habitat is present on all islands, and constitutes a larger proportion of island area on small islands than on large ones. Thus the species of this habitat, and the many endemics among them, form a larger proportion of a small island's total flora than of a large island's total flora.

The Taxon Cycle

As we have argued above, it seems highly probable that island biotas are repeatedly augmented by new immigrants and repeatedly depleted by local extinctions. In other words, it seems highly probable that there is a continual turnover of species-populations and species, although there is considerable disagreement about turnover rates. The prolonged persistence of a species on an island or an archipelago may indicate that a single lineage is persisting, or else that newcomers from the mainland reinforce the original colonists sufficiently often to stave off local extinction. In the former case, given enough time, the enduring species-population is likely to grow in size and (perhaps) expand its initial range and invade more habitats. This ecological diversification may lead to fragmentation of the founding population and thus permit the evolution of new subspecies and species, which are endemic to the island or the archipelago. Finally, since the number of endemics obviously cannot increase indefinitely, some at least will go extinct. This sequence of events is the so-called "taxon cycle."

Several authors have described the cycle, in various contexts, for example, Wilson (1961) in the ant fauna of the Melanesian islands, Greenslade (1968, 1969) in the avifauna and the insect fauna of the Solomon Islands archipelago, and Ricklefs and Cox (1972) in the avifauna of the West Indies.

As a concept, the taxon cycle contains more assumptions than does the simple equilibrium theory of island biogeography. The latter treats each island as an indivisible whole and, if it belongs to an archipelago, disregards its biotic relationships with the other islands. The taxon cycle, however, purports to describe in some detail the stages through which an invading species-population passes within an archipelago, and even within particular islands, and the way in which it becomes adapted to new habitats and undergoes evolutionary diversification in the process. Many plausible sequences can be postulated; studies done so far consist of possible interpretations of what has been observed rather than testable predictions about what will be observed.

A detailed account of the cycle has been given by Ricklefs and Cox (1972). They classified the bird species (excluding hawks and owls) of the Greater and Lesser Antilles into four classes on the basis of their geographic distributions, and they consider that every species (or, at any rate, most of them) will belong to each class in succession, that is, that the members of a class are species in the same stage of a temporal cycle. The stages are as follows (they are shown diagrammatically in Figure 6.9):

Stage I: The species is spreading rapidly through the archipelago and occurs in some or all of the islands; (Figure 6.9 shows a late part of stage I). The populations of the species on the separate islands and on the mainland are undifferentiated from each other.

Stage II: The species has gone extinct on a few small islands. Its populations on the larger islands are clearly differentiated.

Stage III: The population fragments are further reduced in area and have diverged to become separate subspecies or species.

Stage IV: The descendant species' populations have dwindled until each is endemic to a single island. They have become relict species.

Species and species groups having these patterns were recognized by Ricklefs and Cox, and they inferred that the observed patterns repre-

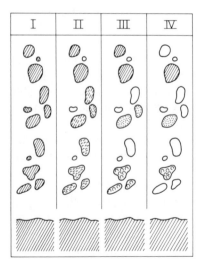

Figure 6.9 The taxon cycle according to Ricklefs and Cox (1972). An idealized representation of the four successive stages. The shapes of the symbols show the way separated populations diverge from one another.

sented stages in a temporal cycle. The possibility that the cycle proceeds in the reverse order, with stage IV coming first, can be ruled out since species found on single islands are true endemics that do not occur on the mainland; thus they cannot be new immigrants but must be relics. Species in stage I are assumed to be in the process of expanding their ranges (or to have just completed the expansion), from a base on the mainland, to cover the whole archipelago. At first they are found only on the islands nearest the mainland, but they spread fast, by island-hopping.

Species in the different stages are further distinguished by their tendency to occur in different habitats; (this distinction is not shown in Figure 6.9). Among birds, typical stage I species are birds of open habitats, often near the coast; typical late-stage species occur in island interiors, in mature forest habitats. This may indicate that populations tend to become restricted to habitats in which the environment remains relatively constant because climatic fluctuations are more buffered than in open, unprotected habitats.

Not all species are assumed to complete the cycle and reach stage IV. Some may remain in an early stage indefinitely, or may undergo reexpansion at some stage and begin the cycle anew. A species that does go right through the cycle to extinction, however, exhibits behavior that can be summarized as follows: it expands its population size, and its geographic range, very rapidly to begin with; thereafter the range fragments and contracts and the separated subpopulations dwindle toward ultimate extinction. Ricklefs and Cox conclude that such a species starts, as an immigrant, with a definite advantage over competing species that immigrated earlier; and that as the species progresses through the cycle it steadily loses its initial advantage. Their explanation for this continuous loss of competitive ability is that the biotic environment into which the species has immigrated becomes "counteradapted" to the new species through evolutionary adjustments; that is, predators and parasites evolve to take advantage of the newcomer, and competitors evolve to match it, so that its initial escape from the predators, parasites, and competitors that it had contended with on the mainland is soon negated.

A far less elaborate explanation of a bird species' progression through the cycle seems (to me) more probable. It accords with the opinion advanced by many workers (for example, Diamond and May, 1976) that land birds seldom cross even narrow arms of the sea voluntarily; and also with the otherwise unexplained fact that the cycle appears to start with a single blanketing invasion of an archipelago by a mainland species with few subsequent followers of the same species. Birds are often transported long distances involuntarily by strong winds and could in this way become involuntary immigrants into several or all of the islands of an archipelago simultaneously. Once there they are trapped, as mainland birds are not, when a climatically "bad" year comes. On the mainland, such a bad year often drives bird populations out of their customary haunts into seldom-visited areas. On an island archipelago this method of escape from harsh conditions is not available; populations can only remain on their islands and endure. A population can improve its chances of survival to some extent by moving into sheltered (especially forested) habitats and this would explain why late-stage populations, approaching extinction, are most frequently those adpted to, and found in, such habitats.

Among insects and plants a similar explanation may apply. The cycle may proceed because, when a local population is drastically reduced by unusually harsh weather conditions, the destruction is selective; the proportion of survivors is greater in sheltered than in exposed habitats.

Thus reduction in population size, contraction of area occupied, restriction to sheltered habitats, and natural selection leading to the differentiation of distinct subpopulations all proceed hand in hand. At the same time, because these events are taking place on islands, which can be invaded from outside only in exceptional circumstances, "good" years cannot repair the destruction caused by bad years.

It is this last point that accounts for the fact that the taxon cycle is an island phenomenon. On the mainland, but not on an island, a gap in a species' range caused by a local climatic "catastrophe" can soon be filled by reinvasion from neighboring areas that were unaffected.

Thus the whole taxon cycle may simply be the effect of sporadically occurring climatic "bad years" on species-populations too isolated for losses to be quickly made good from nearby populations. It is noteworthy that the cycle has been observed in birds and insects, flying and wind-dispersed animals that can be forced into stage I of the cycle by an exceptional storm.

Endemics on Drifting Islands

The taxon cycle has been envisaged as taking place in a temporally constant environment. However, in attempting to interpret the history of a group of species in an archipelago, the geological and climatic history of the archipelago, as well as the evolutionary history of the species, must be considered. The effect of climatic history has already been mentioned (page 194) in connection with the plants of the Galapagos Islands.

The effects of geological history have to be considered when, as sometimes happens, a relict endemic species (or, synonymously, an *epibiotic* species) is found on an island too recently formed for the epibiotic to have evolved there. Several examples among plants have been given by Axelrod (1972), who lists the geologic ages of numerous oceanic islands. Few are older than early Pliocene (10 m.y.) and the majority do not predate the Quaternary (2 or 3 m.y. at most). Many such young islands are the sole places of occurrence of what Axelrod calls "peculiar" epibiotics, "living fossils" that could not have originated on the islands where they are now found. Several of these peculiar epibiotics are genera of trees in families whose modern members are herbaceous. For example, on Socotra Island (at the mouth of the Gulf of Aden), there is an endemic arborescent genus (*Dendrosicyos*) of the Cucurbitaceae (cucumber family). On Juan Fernandez (off the coast of Chile) are three arborescent genera (*Robinsonia,*

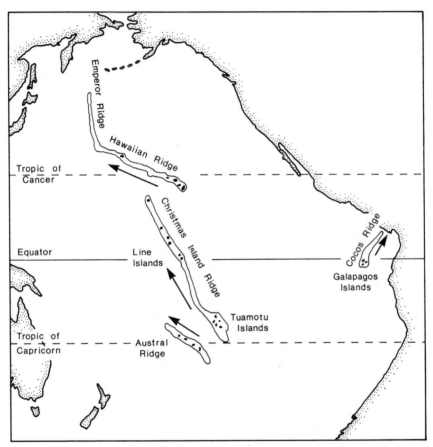

Figure 6.10 Four island chains in the Pacific. Submarine ridges are outlined and the islands are shown black. Arrows show the direction of drift. In each chain, the islands become progressively older in the direction in which the arrow points. (Data from Wilson, 1963b, 1965, and Axelrod, 1972.)

Rhetinodendron, and *Dendroseris*) of the Compositae (Good, 1974) as well as three arborescent species of the genus *Chenopodium* (family Chenopodiaceae). All these taxa are almost certainly older than the islands to which they are now confined.

The probable explanation is that the taxa originated on islands that now have no suitable habitats for them, or even no longer exist except as submarine sea mounts.

Many oceanic archipelagos have the form of chains. Figure 6.10 shows examples in the Pacific. Such a chain is believed to have been formed by a series of volcanic eruptions at a localized hot spot whose position is fixed relative to the earth's center; however, the drift of the crustal plates causes different points on the crust to lie over the volcanic hot spot at different times. As a result, the successively formed islands lie in a chain. Further, the chain itself drifts, as part of its plate, with the oldest islands in the lead and the youngest trailing (Wilson, 1963b, 1965).

As biotic environments, the islands of such a chain are changing continuously. While they age, weathering and (possibly) crustal subsidence cause them to lose height. At the same time, they are drifting across climatic zones. For instance, the oldest islands in the Hawaiian-Emperor chain have drifted from within to outside the tropics; likewise, Christmas Island, now at 2°N, has been rafted to its present position from 15°S where the Tuamotu Islands are now (see Figure 6.10). The average rate of drift is probably 3 to 3.5 cm per year.

Thus the modern climate on an old island is not what it was when the island first formed (quite apart from the concurrent effects of worldwide climatic change). The environmental conditions an island originally offered may be much more closely duplicated, now, on a more recently formed island. These considerations make it likely that some of the epibiotics on young islands have arrived there by island-hopping. As an island on which a new species first evolved drifted away from its area of origin, the species' descendants have maintained themselves in the ancestral environment by shifting to successively younger islands as these were formed.

Chapter Seven

GEOGRAPHICAL ECOLOGY

This chapter considers the geographical distributions of different types of ecological communities, and geographic variation in the ecological processes within communities. The entities whose geographic patterns were studied in earlier chapters were species and other taxonomic groups. Now we consider the geography of larger, ecological entities, such as vegetation types and whole communities of plants or animals. As is well known, collective entities such as sea-bird communities, for instance, have similar ecological properties wherever they occur; but if they are geographically far apart, their member species may be entirely different. This is the topic of Section 1 and is developed further in Section 2.

Some biotic processes can take place on a small, "ecological" scale and also on a large, "geographical" scale. For example, the famous wildlife cycles, the cyclical variations in population numbers in the Canadian lynx, snowshoe hare, and other fur bearers in northern Canada are certainly geographic in scale; but many of the attempts to explain them have been based on theoretical arguments that are thought to apply to such purely ecological phenomena as fluctuations in the size of a flour beetle population in a closed container. Likewise, the single word "competition" is used to describe processes differing enormously in scale. On a large scale is the interaction between, say, herds of large ungulates migrating between the Americas when the Isthmus of Panama became dry land (see page 57), a process whose areal and temporal coverage spanned two continents and perhaps tens of thousands of years. On a small scale is the interaction between two species of yeast competing in a test tube until the outcome is decided within a few days. It is arguable whether a striving for generality justifies an assumption so sweeping as that the processes in the two cases are isomorphic. The subject is discussed in Section 3. Also considered is whether competition on a small (ecological) scale proceeds in the same way everywhere, using *everywhere* in its geographic sense.

Section 4 is a mathematical section. It discusses the problem of examining and mapping geographical variation in the biological properties of a single species. The reason for its inclusion in this chapter is that it deals with a subject that lies somewhere between traditional ecology and traditional biogeography. Ecological studies of a species-population must often cover regions sufficiently extensive for it to be impossible to disregard spatial variation in the population concerned. For instance, particular morphological features may exhibit trends in size or form; or, in polymorphic species, the relative proportions of the different morphs may vary from place to place. Methods of studying such variation exist in enormous numbers and the purpose of the section is to demonstrate two such methods out of the hundreds of possibilities, to give the reader at least a small taste of the subject.

1. REGIONAL COMMUNITIES: RESEMBLANCES AND CONTRASTS

An elementary fact, familiar to all ecologists, is that entirely unrelated organisms can have very similar morphologies if they have become adapted to similar environments. Consider two well-known examples, one from the plant kingdom, the other from the animal. Among plants, arid-land succulents belonging to the families Cactaceae and Euphorbiaceae are not closely related, but because of their morphological adaptation to endure drought, they appear very similar in outward form; saguaro cactuses and candelabra euphorbias are a classic example of this convergent evolution. However, the Cactaceae are confined to the New World, and in the Old their place is taken by succulent Euphorbiaceae, an almost cosmopolitan family. Among animals, the arboreal mammal families Sciuridae and Phalangeridae are not closely related; the former are placentals, the latter marsupials. But in both families, species have evolved with membranes on each side joining the fore and hind limbs which enable them to glide; the Sciuridae include the flying squirrels of North America; the Phalangeridae include the flying phalangers and gliding possums of the Australasian region. In much the same way, convergent evolution can cause ecological communities to resemble each other. In the subsections to follow, several examples are discussed. Examples like these are such a staple ingredient of elementary ecology courses that the student soon comes to accept as commonplace what is, in fact, a remarkable phenomenon.

Often, two communities in similar environments but far apart geographically are found to be very similar in ecological structure. Such matched pairs of communities may be of two kinds. In the first kind, they are composed of matched pairs of closely related species; for instance, in matched forests, the dominant trees in both may belong to one genus, with the difference between geographically separate forests manifesting itself only at the species level. It is not the species, but the communities they compose, that have converged, or at any rate developed in parallel, in different parts of the world. In the second kind of community pair, the component organisms as well as community structure have developed convergently. Thus one finds, in the case of plant communities, for instance, that physiognomically similar vegetation in geographically separate regions may be made up of totally different components; it is in ecological structure only that the separate communities resemble each other, and taxonomically they are entirely distinct.

The Life-Form Spectra of Different Vegetation Types

Studies on the physiognomic resemblance of plant communities that grow in similar climatic environments, even though they are geographically far apart, have a long history. It was in this context that biogeographical data were first treated statistically, by Raunkiaer in his classic book *The Life Forms of Plants and Statistical Plant Geography*.* He developed the concept of the *biological spectrum* of vegetation. A particular vegetation type's spectrum consists of a list of the relative proportions it contains of species with different *life forms*. Plants can be classified into a number of life forms on the basis of the way in which the perennating (or regenerating) organs are protected in adverse seasons. The five chief life forms into which ordinary terrestrial angiosperms are classified are as follows.

Phanerophytes.   Perennial plants whose persisting above-ground stems are of sufficient height to hold the dormant buds well above the ground, out of any sheltered microclimate that may exist near ground level (trees and shrubs).

Chamaephytes.   Perennial plants whose dormant buds, though above the ground, are close enough to it to enjoy a relatively sheltered microclimate; for instance, in cold-temperate climates, snow protects the buds in winter (the dwarf shrubs of heaths and barrens are typical chamaephytes).

Hemicryptophytes.   Perennials whose dormant buds are located at the soil surface (rosette plants are examples).

*Originally published in 1903. An English translation was published by Clarendon Press, Oxford, in 1934.

Geophytes.   Plants whose perennating organs are entirely subterranean (for example, plants with bulbs or corms).

Therophytes.   Annuals; plants in which the only parts to survive through the dormant period are seeds.

To demonstrate the contrasts among the spectra of different kinds of vegetation, it is interesting to consider the proportions of the commonest, and rarest, life forms found in three easily visualized vegetation types (the data are from Dansereau, 1957). In tundra vegetation in the Canadian Arctic, 57% of the species were found to be hemicryptophytes and there were no phanerophytes at all; in desert vegetation (in Cyrenaica), 50% of the species were therophytes (that is, annuals) and 8% were geophytes; in lowland tropical forest (in the Danish West Indies) 74% of the species were phanerophytes and only 1% geophytes. Notice that these percentages are of numbers of species in species-lists, not numbers of individuals, or of cover; also, confidence limits are not given.

An account of these and related topics has been given by Dansereau (1957). In studies on the biological spectra of vegetation, attention has been focused exclusively on the life forms of plants without regard to their taxonomic relationships. For this reason, and also because research on biological spectra appears at present to be (temporarily perhaps) inactive, the subject is not pursued further here. Diagrams of three typical spectra are shown in Figure 7.5, in another context.

Ecologically Matched Vegetation Types

It was remarked above that a pair of "matched" communities (that is, matched in ecological and physiognomic structure) could be of two kinds: either the ecologically equivalent (matched) species in the two communities might be related, or they might be quite unrelated. This is, however, an oversimplification. In matched communities whose matched species are related, obviously the relationship may be of any degree of closeness. It may be very close, as when the matched pairs are pairs of sister species (the immediate descendents of a common ancestor species), or it may be comparatively distant.

To illustrate, consider the forests of temperate North America and Eurasia, whose origins and relationships have been investigated in depth by Kornas. The account below is based on his summary (Kornas, 1972) of this work. Temperate Holarctic forests can be coarsely classified into two main types: the more northern, *boreal,* coniferous forest; and south of it, the broad-leaved, deciduous, *nemoral* forest. Ecologically and floristically these two types of forest are strongly contrasted. At the same time, the boreal forests in the two continents (equivalently, in the Nearctic and Palearctic regions) resemble each other very closely, both ecologically and floristically; the nemoral forests of the two continents also resemble each other but the floristic resemblance is much less close.

We consider the contrasts and resemblances in turn. Table 7.1 summarizes the ecological contrasts, which are striking. Floristically, the contrast between the two forest types is equally striking: there are almost no sister pairs of species with a member in each forest type. This implies that the two communities have separate origins and different histories.

Next consider the floristic relationships, first between the North American (Nearctic) and Eurasian (Palearctic) boreal forests, and then between the respective nemoral forests.

In the boreal forests of the two regions, the majority of the species are of circumboreal range. There is no perceptible difference (or almost none) between their North American and Eurasian representatives. Some examples are: *Goodyera repens* (rattlesnake plantain), *Circaea alpina* (enchanter's nightshade), *Galium triflorum* (bedstraw), *Moneses uniflora* (one-flowered wintergreen), *Pyrola rotundifolia* (round-leaved wintergreen), *Arctostaphylos uva-ursi* (bearberry), *Empetrum nigrum* (crowberry), and all the mosses and lichens which are important components of the ground layer in these forests. Less numerous than the circumboreal

Table 7.1 Ecological contrasts between the Holarctic Boreal forest and the Holarctic Nemoral forest

Ecological characteristic	Boreal (coniferous) forest	Nemoral (broad-leaved deciduous) forest
Number of species	Few	Many
Soil	Poor, podsolized	Relatively rich brown forest soils
Litter	Acid	Neutral
Canopy layers	One	Two or three
Large shrubs	Few	Many
Field layer	Low and dwarf shrubs	Herbs
Mosses and lichens	Abundant	Sparse
Light regime below the canopy	Constant through the year because trees evergreen	Light in spring, dark in summer, because trees deciduous

species are taxa that have become slightly differentiated, into closely similar matched pairs of subspecies or species. Some representative examples, with their popular names, are listed in Table 7.2. Notice that a complete list would include all the trees of the canopy; no tree species are common to both regions. There are very few species in the boreal forests of either continent that are not represented by the same or a closely related species in the other; such few "unmatched" species as there are are "strays" that chance to have invaded from other community types.

The floristic resemblance between North American and Eurasian nemoral (broad-leaved deciduous) forests is not nearly so close as that between their boreal forests. There are far fewer species common to both and these are chiefly ferns, for example, *Dryopteris spinulosa, Athyrium filix-femina*, and *Thelypteris phegopteris*. But there are numerous matched pairs of species, of which a few representatives are listed in Table 7.3. The morphological resemblance between the members of each pair in Table 7.3 is not so close as for pairs in Table 7.2. Figure 7.1 shows drawings of the leaves of four of the paired tree species listed in Table 7.3 (an analogous figure for the pairs in Table 7.2 has not been given as it would show only pairs of scarcely distinguishable "twins"). It is noteworthy that among these broad-leaved trees, Nearctic trees tend to have larger leaves than their Palearctic counterparts.

Table 7.2   Matched pairs of species in the boreal forests of North America and Eurasia (adapted from Kornas, 1972)

North American species	Eurasian species	Popular name of the genus
Trees		
Pinus banksiana	*P. sylvestris*	Pine
Abies balsamea	*A. sibirica*	Fir
Picea glauca	*P. excelsa*	Spruce
Sorbus americana	*S. aucuparia*	Mountain-ash
Angiosperm herbs		
Maianthemum canadense	*M. bifolium*	Wild lily-of-the-valley
Oxalis montana[a]	*O. acetosella*	Wood-sorrel
Linnaea americana[a]	*L. borealis*	Twinflower
Fern		
Polypodium virginianum	*P. vulgare*	Polypody

[a] These species are treated as conspecific subspecies, not as separate species, by Kornas (1972).

Table 7.3 Matched pairs of species in the nemoral forests of North America and Eurasia (from Kornas, 1972)

North American species	Eurasian species	Popular name of the genus
Trees		
Acer saccharum	A. platanoides	Maple
Fagus grandifolia	F. silvatica	Beech
Fraxinus nigra	F. excelsior	Ash
Quercus alba	Q. robur	Oak
Tilia americana	T. cordata	Basswood or Lime
Ulmus americana	U. laevis	Elm
Shrub		
Viburnum trilobum	V. Opulus	Viburnum
Herbs		
Actaea rubra	A. spicata	Baneberry
Asarum canadense	A. europaeum	Wild ginger
Circaea quadrisulcata	C. lutetiana	Enchanter's nightshade
Sanicula trifoliata	S. europaea	Black snakeroot or Sanicle

The exceedingly close resemblance between North American and Eurasian boreal forests and the much less close resemblance between their nemoral forests is presumably due to their different histories. Kornas (1972) offers the following as a possible outline. Coniferous forests probably originated in the mountains of the temperate zone early in the Tertiary and spread into the lowlands as the world climate cooled. They finally coalesced into one unbroken circumpolar forest, most of which was destroyed by the Pleistocene ice sheets, although it is possible that a narrow but continuous strip persisted around the periphery of the ice even at the time of maximum glaciation. Thus the separate blocks of forest that now exist became isolated from each other no earlier than the Pleistocene and the gaps separating them are still narrow.

The separation into isolated blocks of the nemoral forest is much older; it probably occurred no later than the Pliocene and perhaps earlier. Before that, when the world climate was warmer, the nemoral forest formed a continuous, circumpolar vegetation zone as the boreal forest does now (see Chapter 4, page 131). This so-called Arctotertiary forest was more uniform floristically and richer in species than its three surviving blocks or, as Kornas calls them, refugia, in, respectively, eastern North America, eastern Asia, and southeastern Europe. The original Arctotertiary forest was

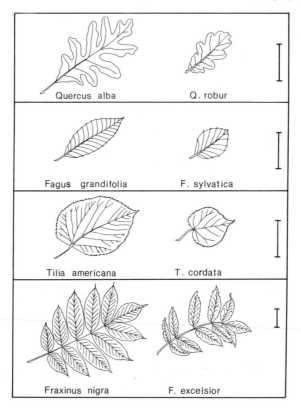

Figure 7.1 The leaves of four congeneric pairs of tree species that constitute "ecological replacement pairs." *Left:* North American species. *Right:* Eurasian species. All scale marks represent 5 cm.

forced southward ahead of the expanding boreal forest, and became separated into the three surviving blocks because of the worldwide cooling of the climate in the late Tertiary.

Related species in the nemoral forests of different geographic areas have therefore been diverging evolutionarily for a longer time than have those of the boreal forests, and this is assumed to account for their more pronounced morphological differences (if this is true, it is an example of phyletic gradualism; see Chapter 3, page 67). However, Kornas considers that such divergence as has occurred, in taxa in both types of forest, has been morphological only, not ecological. The reason for supposing this is that the members of matched pairs of species invariably occupy matching communities in their respective continents; that is, there are no known cases whatever of a species-pair having its American member in the boreal forest and its Eurasian member in the nemoral forest or vice versa. "The conclusion is that the ecological constitution of such taxa must be very rigid" (Kornas, 1972) and may have persisted unchanged for 10 to 15 m.y. (that is, since the Pliocene or Miocene) while minor morphological differentiation was taking place. The presumption is that some morphological characters are ecologically neutral and nonadaptive, but that they have, nevertheless, become fixed in their respective new species; these conclusions will not go unchallenged, and invite further research.

Another vegetation type that is determined by climate, and which exhibits considerable ecological and physiognomic uniformity, is that of regions with a so-called Mediterranean climate. The climate is characterized by warm, moist winters and hot, dry summers. This climate, and (on land) its associated vegetation, is unlikely ever to have formed a single circumglobal zone. Even so, there is a very strong floristic resemblance between the plant communities of two widely separated areas in the northern hemisphere with Mediterranean vegetation, namely Mediterranean Europe and California. At the same time, the vegetation in these two areas is entirely different floristically from the vegetation in parts of Australia with a Mediterranean climate (Specht, 1969) although the ecological and physiognomic resemblance is close. On fertile soils, the climax vegetation is open savannah woodland of fairly small, evergreen sclerophyllous* trees.

In Mediterranean Europe and California, the dominant tree genus is *Quercus* (oak). Besides the oaks, there is a long list (Raven and Axelrod, 1974) of other genera of woody plants common to both regions, among them *Acer, Arbutus, Cercis, Clematis, Cupressus, Fraxinus, Juniperus, Lonicera, Pinus, Platanus, Populus, Prunus, Rhamnus, Rhus, Rosa, Rubus, Smilax, Viburnum,* and *Vitis.* This large number of shared genera suggests that the separation of Europe and North America may have been more recent than is supposed from other evidence; another possibility is that the extinction rate is low in vegetation of this kind.

Those parts of Australia with a Mediterranean climate also have Mediterranean vegetation—open sclerophyllous woodland—but all the species in it are unrelated to their northern hemisphere counterparts. The dominant genera are *Eucalyptus* and *Casuarina.*

The ecological resemblances and floristic contrasts between northern and southern hemisphere vegetation are treated further below, in the

---

*Sclerophylls, which are characteristic of Mediterranean vegetation, are plants whose leaves are thick, tough, leathery, and often rather small, and which persist for two or three years before being shed. From the Greek "scleros," hard, and "phyllon," leaf.

context of fire-controlled vegetation. Before turning to this topic, however, another example, this one zoological, of closely matched communities is described.

Matched Sea-Bird Communities

The possession of apparently identical ecological structures by two or
more geographically separate communities is not limited to plant communities. We now consider a parallel situation among animals, specifically, sea birds.

Bourne (1963) made detailed studies on the structure of two sea-bird
communities. The communities live, and feed, in areas of upwelling water
in two widely separated oceans: in the Atlantic, off the west African
coast, and in the Indian Ocean off the south coast of Arabia.

Each community has about 30 species belonging to six families. Some
are resident and some migrant. Some species occur in both geographic
areas, but others form matched pairs: a particular ecological "role" is
taken by one species in the Atlantic and by another, closely related,
species in the Indian Ocean.* Within either one of the communities, each
species is unique in respect of at least one of the following characteristics:
nature of food; size of food item; method of capturing it; season when the
bird is present; distance from shore of the bird's feeding area.

Apparently, therefore, no two species within one community have
identical abilities and requirements. For example, each region has three
different shearwater species (Procellariidae) but, in each region, one is
small, one medium-sized, and one large, and their food is differentiated
correspondingly. Table 7.4 summarizes the comparison. The large number of matched pairs of species—or ecological replacement pairs as
they may be called—is strong evidence for the "structured," as opposed
to fortuitous, nature of these bird assemblages.

Parallel Evolution in Fire-Dependent, and Other, Communities

We have already considered the close similarity, in ecology and physiog-
nomy, of vegetation in closely similar climates. The component species
may be quite unrelated, as are the plants in semidesert vegetation in
America and Australia, for instance, but they often resemble each other in
many obviously adaptive† characters, and can be said to exhibit parallel
evolution.

*To say, as many would, that such species occupy the same *niches* in their respective
communities is to use that ambiguous word in its Eltonian sense. See Chapter 3, page 94n.
†Attempts to find adaptive "explanations" for the observed characteristics of organisms can

Table 7.4 Composition of two sea-bird communities, one in the Atlantic, the other in the Indian Ocean (adapted from Bourne, 1963)

Family	Food	Species common to both regions	"Paired" species[a]
Phalaropidae (phalaropes)	Zooplankton (collected by birds swimming)	2	—
Hydrobatidae (storm petrels)	Zooplankton (collected by birds in flight)	3	(3)
Sternidae (terns)	Larger zooplankton and small fish, from the air	4	4
Laridae (gulls)	Larger fish	3	2
Stercorariidae (jaegers)	Food stolen from gulls and cormorants	1	—
Procellariidae (shearwaters and petrels)	Fish and some cephalopods (shearwaters); mostly cephalopods (petrels)	—	4[b] (1)
Sulidae (gannets and boobies)	Fish, from the air	1	1
Phalacrocoracidae (cormorants)	Fish, by dives from the surface	—	1
Fregatidae (man-o'-war birds)	Food stolen from smaller birds	—	1
Phaethontidae (tropic birds)	Fish and cephalopods	1	—

[a]Unbracketed numbers show the number of replacement pairs having one member in each ocean. Bracketed numbers show Atlantic species not represented by a replacement species in the Indian Ocean.
[b]Three species of shearwater and one of petrel.

Climate is not the only factor guiding parallel evolution in vegetation, however. Another exceedingly important factor is fire. It is not independent of climate, or course: fires are far more frequent in dry than in wet climates. But it appears likely that many of the selectively advanta-

---

easily be pushed too far, and often are by enthusiasts. In considering parallel evolution in response to parallel environmental stresses, Seddon's (1974) warning is worth bearing in mind: " 'Adaptive' is a theory-laden term . . . There is an ineradicable element of speculation" in claims that a character does now, or has in the past, given its possessors some advantage over competitors.

geous characters possessed by plants that thrive in dry climates are adaptations to fire rather than to drought per se.

Mutch (1970) has argued that in very dry climates, fire replaces bacterial and fungal decomposition as the agent degrading litter and permitting the recycling of its mineral content. Fire is therefore a *necessary* occurrence for the maintenance of such vegetation. Without fairly frequent fires, it could not persist. Because of this, Mutch considers that many of the woody plants in fire-dependent (or fire-perpetuated) vegetation have been selected not merely to withstand fire but to promote it. They are selected for high flammability, causing fires to be more frequent, and hotter, than in vegetation whose plants are not so adapted.

It is interesting to list the characters that promote fires and that enable plants to survive it, and to consider how unrelated plant species in different parts of the world have evolved these characters in parallel.

1. Litter in fire-dependent forest is more abundant and more flammable than in the forests of wet climates. Both the litter and the living plants in fire-dependent vegetation have low moisture content and high content of flammable oils and resins. In combustion tests, Mutch (1970) showed that the litter in fire-dependent forests, compared with that in fire-independent forests, burned faster and more completely, and yielded more heat. For example, open *Eucalyptus* forest in Australia (Victoria), and open forests dominated by *Pinus ponderosa* in the arid American west (Idaho) resemble each other in their high flammability and differ from geographically close but climatically different forests. The contrasted forests, monsoon forest in Australia and wet-temperate hardwood forest in America, are not fire-dependent. Their trees and litter are much less flammable and natural fires are rare.

2. Often the individual trees in fire-dependent vegetation are adapted to recover quickly after a fire. They can produce new shoots from protected underground parts such as rhizomes or woody rootstocks (lignotubers, found in many *Eucalyptus* species).

3. The seeds of many fire-dependent trees are retained in fire-resistant woody fruits and released only some time after a fire, when the ash has had time to cool. The seeds germinate readily in a bed of ashes, and the seedlings are fast-growing and intolerant of shade. They are therefore at a competitive advantage vis-à-vis shade-tolerant trees in newly burned-over land.

The distinction between drought-adapted and fire-adapted characters is hard to draw. When apparently drought-resistant (xerophytic) plants dominate in frequently burned areas where drought is not particularly severe, it may be fire, not drought that is responsible for their dominance.

Seddon (1974) has described "wet sclerophyll forest" in parts of eastern Australia in which there is no seasonal drought but where, all the same, fires are common; the dominant trees—tall eucalypts, 30m or more in height—are sclerophyllous and form an open-canopied woodland; the trees might be accounted xerophytic were it not for the fact that the undergrowth consists of lush mesophytic vegetation with many tree ferns. There are two possible explanations for this apparent mismatch between canopy and under story: (1) Perhaps the sclerophyllous eucalypts were selected for drought-resistance in the past and now, although climatic change may have made the adaptation needless, there has not yet been time for them to evolve better-adapted forms, or to be excluded by better-adapted competitors. (2) Perhaps the sclerophyllous eucalypts are selectively favored over rain forest species because fires are frequent. Fires affect the soil in a way that promotes rapid leaching of nutrients, and sclerophylls are known to be better adapted than rain-forest species to infertile soils. The cycle of cause and effect may thus be of the form:

$$\text{Fires} \quad \begin{array}{c} \nearrow \\ \leftarrow \\ \longrightarrow \end{array} \quad \begin{array}{c} \text{Sclerophyllous forest} \\ \uparrow \\ \text{Infertile soil} \end{array}$$

where arrows imply causation.

Differences in the frequency of fires may also account for more subtle differences in vegetation than that between rain forest and arid-land sclerophyllous vegetation. Thus in the boreal forests of eastern North America, the composition of the tree communities in different areas is governed by susceptibility of the area to fires (Rowe and Scotter, 1973). Fires tend to be most frequent where the topography favors quick drying of the vegetation following rain; convex ground surface, southern exposure, and porous soil are all fire promoters. The tree communities in such areas (in the boreal forest zone) consist of species that are quick to invade newly burned land, for example, *Pinus banksiana* (jack pine), *Populus tremuloides* (aspen), *Betula papyrifera* (paper birch), *Picea mariana* (black spruce), and *Larix laricina* (larch, tamarack). *Picea glauca* (white spruce) is a comparatively slow colonizer and *Abies balsamea* (balsam fir) is slower still. Therefore the last two species cannot establish themselves where fires are frequent; they are dominant chiefly in the wettest, least fire-prone areas, for example, in parts of Newfoundland. It is worth noting that balsam fir, which is most abundant in relatively firefree sites, has cones that disintegrate, and scatter their seeds when mature; jack pine has cones that stay tightly closed until fire causes them to open and release their seeds.

There is considerable geographic variation in fire susceptibility in the northern forests, and also considerable disagreement over the effects of fire. According to Rowe and Scotter (1973), fire can convert forest to tundra: forest is the climax vegetation, but repeated fires can perpetuate a successional stage, namely tundra. The converse view is held by Strang (1972) who considers that, in the northern part of the boreal forest zone, spruce forests are successional and are perpetuated by fire; and that if fire is excluded, a tundra of dwarf shrubs and lichens succeeds it.

Went (1971) has described several cases of what appears to be parallel evolution in plants, but with two very curious properties. The common character that has presumably evolved independently in several unrelated species has no discernible adaptive value. And it has evolved in only one (or at most two) geographic regions in the world and is not, there, restricted to any one particular environment.

We consider three of Went's eight examples.

1. In New Zealand, but nowhere else in the world, shrubs with a *divaricate* form of branching are common. This growth habit is very distinctive: the shrubs are small and have intricately interlaced branches that emerge at right angles to the main stem, separated by long internodes; the leaves are small. This form of branching is found in 50 species in 21 families (including Violaceae, Rutaceae, Malvaceae, Polygonaceae, Rubiaceae, Araliaceae, Cornaceae, Moraceae). Each species has congeners that are not divaricate and divaricateness has no evident adaptive value. The divaricate shrubs are so similar to each other as to be almost indistinguishable when not in flower.

2. In a great variety of environments in Australia, and nowhere else, plants with a eucalypt leaf type are found. They occur in 12 families, among them the Myrtaceae (to which the genus *Eucalyptus* belongs), Proteaceae, Bignoniaceae, Leguminosae, Apocynaceae, Polygonaceae, Labiatae, and Compositae. The distinctive characteristic of these leaves is that there is no difference between their dorsal and ventral sides; palisade parenchyma occurs on both sides. The leaves are usually lanceolate in shape, tapered at both ends, and oriented vertically, either hanging or erect. It would not be difficult to imagine advantages bestowed by this type of leaf on a plant that bears them if such plants were limited to environments of one particular kind; but they are not so limited.

3. Trees and shrubs whose leaves turn scarlet, crimson, and purple in the fall (as, for example, *Acer rubrum*, red maple, and *Quercus coccinea*, scarlet oak) are nearly all natives of northeastern North America or northeastern Asia. The characteristic is rarely found in plants native to other regions.

These examples, and the other five described by Went, are cases of parallel evolution that appear to confer no selective advantage on the plants concerned and that are very localized geographically. Went's own hypothesis is that it is due to nonsexual transfer of genetic material, perhaps by insect vectors, between unrelated plants. Naturally occurring "interfamilial chromosome transfer" will be difficult to demonstrate, if it occurs, and it is hard to accept merely on the basis of circumstantial evidence, although it cannot be flatly ruled out. The problem remains to be solved.

2. LATITUDINAL AND ALTITUDINAL ZONATION AND THE ELEMENTS OF REGIONAL BIOTAS

This section discusses three interrelated topics: the latitudinal and altitudinal zonation of biotas, and the origin of the biotic elements that combine to form the biota of a region (the term *element* is defined in context, in the discussion below).

The Zonation of Vegetation

The latitudinal zonation of terrestrial vegetation, especially in the land hemisphere of the world—the northern hemisphere—is well known, so well known that it is taken for granted. Figure 7.2 shows the zonation in comparable strips, between two meridians of longitude 5° apart, in North America and Eurasia. It is seen that the ordering of the zones, from arctic tundra in the extreme north, south to the nemoral forest zone is identical in the two continents. South of the 40°N parallel in the North American strip, the clarity of the zonation disappears owing to "interference" from a northeast to southwest trending mountain range, the Alleghenies; but woodland-steppe and steppe (zones 7 and 8 in the mapped Eurasian strip) are present, of course, farther west in North America, the latter under the name of "prairie."

The figure also emphasizes that although the ordering of the zones is identical in the two landmasses, the pattern as a whole is located much farther south in eastern North America than in Eurasia. This is a consequence of the cooling effect of Hudson Bay (see Chapter 4, page 124). The locations of the zone boundaries are governed by climate. In general terms, they parallel the winter and summer isotherms as may be seen by comparing them with standard climatological maps. However the most important single factor appears to be the duration of the growing season, the period in which the mean daily temperature exceeds 6°C; winter cold,

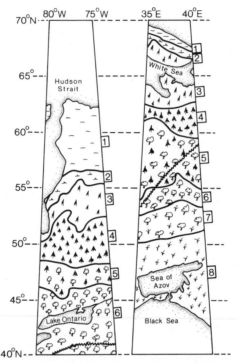

Figure 7.2 Latitudinal zonation of vegetation in North America and Eurasia. The zones (labeled with numbers in boxes on the right of each map) are: 1, Arctic Tundra; 2, Woodland Tundra; 3, Subarctic; 4, Boreal; 5, Boreo-Nemoral Transition; 6, Nemoral; 7, Woodland-Steppe; 8, Steppe. (Adapted from Sjörs, 1963.) The zigzag line across each map shows the southern limit of the ice at the last glacial maximum. (After Flint, 1957.)

and the presence of permafrost, appear to be comparatively unimportant (Sjörs, 1963).

The most striking feature of the maps, after allowing for the relative displacement of their patterns as a result of climatic contrasts, is their similarity. The sequence of ecological communities encountered along a meridian is the same in both continents. This fact is worth contemplating in conjunction with the fact that the vegetation of the nemoral forest zone and of all zones north of it has existed for only 10,000 years (approximately) since the melting of the Wisconsin and Würm ice sheets. The matching zones or, equivalently, the matching plant communities, have come into existence as a result of the northward dispersal into newly icefree land of plants that had survived the Pleistocene glaciations in refugia south of the ice. The species constituting the communities of a given zone in the two continents are not the same (see page 206) but the ecological structures of the matched communities are apparently identical. It is not known when these community types first appeared as community types, and whether their origin was contemporaneous with, or later than, the evolution of the component species. But the actual communities of each type appear to have migrated as entities. It seems most unlikely that, if the component species had all migrated northward independently of each other, they would have combined, as they have, to form such closely similar communities in the two continents.

As to whether the zones are clearly distinct or merge into each other, no single answer is possible. All degrees of distinctness are found, in different places. Observations on latitudinal zonation, therefore, cast no light on the question of whether, in general, vegetation on a gradient forms discrete communities or a continuum (see McIntosh, 1967; Langford and Buell, 1969; Pielou, 1977). Probably there is no one general answer and the question itself is meaningless.

We come now to altitudinal zonation. As is well known, the biotic zones traversed as one ascends a mountain at low latitude simulate the zones that would be traversed if, starting from the same point, one journeyed poleward at sea level. The resemblance of high altitude alpine vegetation (defined as that of the zone between timberline and the permanent snowline) to arctic tundra is striking. Even when there is no floristic resemblance, as there is not between Palearctic tundra and alpine vegetation in the Andes, for example, there is close physiognomic resemblance; equivalently, the biological spectra of the plants' life forms (see page 203) are similar.

Often there is floristic resemblance as well. Many plant species that are widespread at high latitudes have extensions of range into low latitudes where a north-south mountain chain provides a "finger" of suitable habitat. This is particularly well shown in North America where arctic plants can extend their ranges southward along the Appalachian mountain chain in the east and along the Rockies and parallel mountain chains in the west. A typical example is *Solidago multiradiata* (Northern goldenrod). It is widespread in subarctic Canada, Alaska, and northeastern Siberia. At lower latitudes, it is found only at high altitudes: in eastern North America in the Shickshock mountains of the Gaspé Peninsula (Québec) and in the highlands of Cape Breton Island (Nova Scotia); in western North America, in the alpine zones of mountains in Colorado, Arizona, and California.

Likewise at lower altitudes, below timberline, many tree species are found whose zones on mountain slopes become steadily higher with decreasing latitude. For instance, *Pinus lambertiana* (sugar pine) occurs at altitudes between 520 m and 1100 m in Washington State and between

2400 m and 3100 m in Baja California (Mirov, 1967). The form of such geographic ranges shows that it is temperature-related factors—for instance, length of the growing season or of the frostfree period—that have the strongest effect on plant distribution. The great contrast between high and low latitudes in the number of hours of daylight in a summer day does not appear to influence the spread of plant species.

Ideally, one might expect to find perfect correlation between two lists of species: a list giving the order, from high to low altitude, of the zones on a mountainside dominated by different plant species; and a list giving the order, from high to low latitude, of the poleward range limits of the same species. Such precise agreement, though easy to visualize, is in fact seldom approached. It might (possibly) be reasonable to expect a high correlation if the widths of the altitudinal ranges of the species concerned were all of roughly the same magnitude and likewise for their latitudinal ranges. In fact, species vary enormously in the extent (vertical and horizontal) of their ranges.

This suggests that it would not be surprising to find that the magnitude of a species' latitudinal range is correlated with that of its altitudinal range. Such correlation is indeed found in the pines of western North America. The rank correlation coefficient (Spearman's coefficient) between the magnitudes of the altitudinal and latitudinal ranges of 14 of these species* is found to be (using data from Mirov, 1967) $r_s = 0.770$ for which $P < 0.006$ (one-tailed).

The interpretation to attach to this finding is not clear, however. The one that first springs to mind, namely that both ranges reflect the ecological amplitude (environmental tolerance range) of a species, is not necessarily the correct explanation. One of two other possible explanations may, in particular cases, be the true one.

The first possibility is that a species approaching extinction, and whose total numbers are dwindling, has been driven, by more successful competitors, into the "marginal" habitats of high mountains. Its range would then consist of disjunct high altitude "islands." When it had gone extinct in all but one of these, its altitudinal and latitudinal ranges would both be small and they would not necessarily reflect its ecological amplitude.

The second possible explanation stems from two well-known facts. An isolated species-population on a mountain top is especially likely to become differentiated from other populations of its species and to be the progenitor of a new species; and a new species, until it has had time to

*The species of *Pinus* considered are *albicaulis, flexilis, lambertiana, monticola, aristata, monophylla, edulis, quadrifolia, sabiniana, coulteri, jeffreyi, ponderosa, contorta,* and *attenuata*. Five other species recorded by Mirov for the region were disregarded because of their rarity and spotty distributions.

disperse, must obviously have a restricted geographic, and hence latitudinal, range regardless of its ecological amplitude. Thus in young species, limited altitudinal and latitudinal ranges tend to be found together.

These two possible explanations amount to supposing that species with small ranges are chiefly mountain endemics, which are either relict endemics (the first explanation) or new species (the second). Endemic species are, indeed, a characteristic feature of mountain biotas as they are of islands in the sea (see Chapter 6, page 192) and may be of either of these two radically different kinds. However, to quote Good (1974), "There is no ready means, except in very rare cases, of knowing whether an endemic species is new or old."

Attempts to trace the history of mountain vegetation, and to correlate altitudinal and latitudinal zonation, are fraught with difficulty because so many processes, abiotic and biotic, have been going on simultaneously through geological time. To consider abiotic change first: most of the high mountains of the modern world were rising throughout the Tertiary and some did not even begin to form until comparatively late in that period; for example, in the American west, the uplift of the Sierra Nevada began in the Miocene, and of the Cascades in the late Pliocene, only 4 m.y. ago (see Stokes, 1973). Simultaneously, the climate of the whole world was becoming cooler, at a rate that no doubt varied from place to place; the crustal plates of the earth, and with them the continents, were shifting their positions in latitude and relative to each other. Hence the climatic history of a given point on the ground is exceedingly hard to unravel, either with or without the aid of fossil evidence.

Biotic changes were proceeding simultaneously at various rates. It will be recalled (see Chapter 2, page 41) that speciation is much faster in some taxonomic groups than in others—faster in angiosperms than in gymnosperms, for example—and that improved plant dispersal mechanisms were evolving also. Besides evolving, species-populations were expanding, contracting, and migrating. And different species vary enormously in their ecological amplitudes, some being found in different environments in different parts of their ranges; for example, *Picea mariana* (black spruce) grows in sphagnum bogs in the southern part of its range, and in stony, well-drained soils in the north (Hosie, 1969); *Pinus sylvestris* (Scots pine) behaves in much the same way, being a xerophyte in the northern part of its range and a mesophyte farther south (Mirov, 1967); (no doubt there are marked genetic differences between the widely separated species-populations).

Finally, whatever sorting into zones occurred in northern latitudes in the Tertiary was destroyed by the Pleistocene ice sheets and had to be reconstituted during the Holocene. The wonder is not that zone patterns are confusing, but that they are interpretable at all. It is interesting, though it may be unprofitable, to speculate on whether, in a temporally constant world (climatically and geomorphologically), vegetation would exhibit more or less latitudinal zonation than it does in fact. Does the differential migration of species in response to change reinforce or obscure the zonation caused by the zoned climates on a rotating sphere warmed by a small, distant sun?

The Elements Composing a Biota

When one examines the biota of a region, it is usually found to contain portions of several zone biotas (of both latitudinal and altitudinal zones). These various ingredients of a regional biota constitute its elements. A faunistic or floristic *element* is defined as a group of organisms that share the same evolutionary and migrational history.

Thus in much of Canada, as an example, the major vegetation zones (those shown in Figure 7.2) have, on balance, been migrating northward since the disappearance of the Wisconsin ice sheet. As a result, the flora of a particular region, although dominated by the vegetation of its zone, may have, mixed with it, some survivors from the vegetation of a (now) more northerly zone that chance to have been left behind in the migration. An example (mentioned already in Chapter 4, page 121) is the arctic tundra plant *Dryas integrifolia* which still persists in pockets of suitable habitat on the north shore of Lake Superior, surrounded by boreal forest.

To illustrate the way in which several separate elements combine to form a regional flora, consider the flora of Nova Scotia, a relatively small, distinct region whose flora has been investigated in detail by Roland and Smith (1969). The region lies entirely within the boreo-nemoral transition zone (Sjörs, 1963) and plants of this zone are therefore the biggest element and form the "background" flora. However, several other elements are present also, and have limited ranges within the region. Some are restricted to unusual habitats and others occur more generally. We consider two especially distinctive elements as examples (see Figure 7.3).

1. The arctic-alpine element consists of plants of the modern arctic tundra zone that have been left behind in the post-Pleistocene migration of vegetation northward. These plants have managed to persist, in spite of competition from more recent immigrants, only in such cool habitats as shaded ravines and fog-shrouded coastal cliffs in certain parts of the region. Examples are *Dryopteris fragrans* (fragrant fern), *Asplenium viride* (green spleenwort) (see Figure 7.3), and *Solidago multiradiata* (northern goldenrod) as already mentioned on page 121.

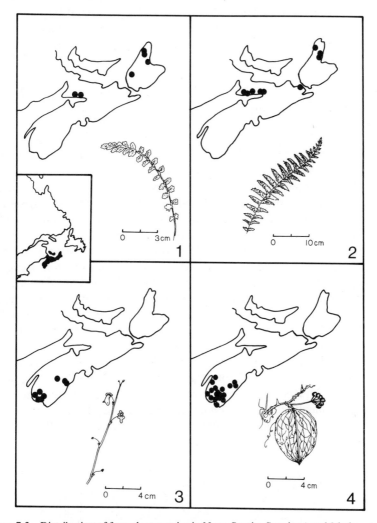

Figure 7.3 Distribution of four plant species in Nova Scotia. Species 1 and 2 belong to the Arctic-Alpine floral element; species 3 and 4 belong to the Atlantic Coastal Plain floral element. 1, *Asplenium viride*; 2, *Dryopteris fragrans*; 3, *Utricularia subulata*; 4, *Smilax rotundifolia*.

2. The coastal plain element consists of plants of southern origin that migrated northeast along the Atlantic coastal plain when it was wider than it is now because of the lower sea level in periods with larger ice sheets. These plants invaded Nova Scotia at its southwestern extremity, when the land became icefree, and that is where they occur now; their ranges

within the region being considered is thus very restricted but outside the region they are found along the Atlantic coastal plain as far as the Gulf of Mexico; examples are *Utricularia subulata* (a bladderwort) and *Smilax rotundifolia* (cat brier) (see Figure 7.3).

It should be noticed that what, to an ecologist, constitutes a single community, may contain species belonging to several different elements. In Nova Scotia seashore communities, for instance, are found plants belonging both to the coastal plain element (whose origin is described above) and to the amphi-Atlantic element; the latter element consists of presumed descendents of a single circum-North Atlantic coastal plain flora that are now found only in eastern North America and Europe. Typical representatives of these two elements are, from the coastal plain element, *Spartina patens* (salt hay), *Limonium nashii* (sea lavender), *Solidago sempervirens* (seaside goldenrod), and *Euphorbia polygonifolia* (seaside spurge) (and see page 115); from the amphi-Atlantic element, *Carex paleacea* (a sedge) and *Ligusticum scothicum* (scotch lovage).

The proportions in a plant community of the different floral elements in its makeup can be represented by a *distributional spectrum*. Some examples of such spectra (from Kornas, 1972) are shown diagrammatically in Figure 7.4. The elements in this case are more inclusive (more broadly defined) than those discussed above. The pie diagrams show the composition of the field layer (herbs, grasses, and dwarf shrubs) in three forests. The first two describe coniferous forests in British Columbia and Quebec; as may be seen, the relative proportions of the different elements in the two geographically distant communities are similar. The element peculiar to each—a western North American and an eastern North American element respectively—amount to less than one-sixth of the total in either

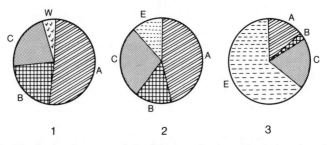

| 1 | 2 | 3 |

Figure 7.4 Distributional spectra of the field layer in three North American forests. 1, Coniferous forest in British Columbia; 2, coniferous forest in Quebec; 3, deciduous forest in Quebec. The floral elements are: A, Circumboreal; B, Asiatic-American; C, North American Transcontinental; E, Eastern North American; W, Western North American. (Redrawn from Kornas, 1972.)

community and about half the species in each are circumboreal. The third pie diagram describes a deciduous, maple-hickory forest in Quebec, geographically close to the Quebec coniferous forest but conspicuously different from it in composition; the eastern North American element accounts for two-thirds of the community and the circumboreal element for only one-sixth. (Compare these diagrams with the descriptions, on pages 205 to 206, of the boreal and nemoral forest zones.)

*3. INTERSPECIFIC COMPETITION: BIOGEOGRAPHIC CONSIDERATIONS†

One of the most widespread of ecological phenomena is interspecific competition. The literature dealing with it is vast. Besides particular studies on all manner of organisms, there are numerous more general discussions on how competition should be defined and on how it manifests itself.

One of the most carefully studied of the putative results of competition is competitive exclusion (Pielou, 1975). The evidence that it takes place, at least in small communities, seems compelling. At the same time, it is known that related species tend to have similar geographic ranges. Thus if a genus, or family, is endemic to a limited geographic region, all its member species necessarily occur together in that region. For example, among mammals, all the approximately 16 species (in 6 genera) of the Lemuridae (lemurs) are confined to Madagascar and the Comoro Islands (Walker, 1968). Likewise among angiosperms, the enormous genus *Eucalyptus* with several hundred species (the "exact" number depends on the taxonomic treatment) is almost entirely restricted to Australia (Good, 1974). This is not to say that related species occurring together in a single region do not exclude each other from particular patches of habitat; although their ranges may appear to overlap if shown without detail on a small-scale map, an accurate large-scale map may show that there is little or no overlap of range on an "ecological" or microgeographical scale; instead, the several ranges may form an intricate mosaic. But it does suggest that interspecific competition may be more of an ecological than a biogeographic phenomenon.

This statement is, of course, too vague to be testable and must be reformulated before we consider whether it should be accepted as true or rejected as false. We now consider how this may be done.

---

†The section is starred because it is, in part, mildly mathematical. The mathematical content is more elementary than in other starred sections, however.

Testing for Competitive Exclusion on a Geographic Scale

The question to be answered is this: if we define any species' geographic range as the smallest convex hull* enclosing all points where the species is known to occur, do the ranges of a group of related species tend to overlap each other markedly (because of their common ancestry) or to overlap each other but slightly if at all (because of competitive exclusion)? There is a third possibility, of course: the species' ranges may be located independently of each other and their overlap be neither "marked" nor "slight." The three degrees of overlap can be described by the respective adjectives *marked, random,* and *slight;* more precise terms have yet to be coined.

Unless the overlap is slight, we can say that there is no evidence of competitive exclusion (in the taxonomic group under consideration) on a biogeographic scale. Such a statement would imply nothing whatever about the occurrence or nonoccurrence in the group of competitive exclusion on an ecological scale—in other words, the exclusion of some species by others from particular habitat patches within their common geographic range—which is a different problem altogether.

A simple, direct attack on the problem is easiest if we examine the pattern of the ranges of species belonging to two-species genera (in the class or phylum being considered). Suppose data are available on the ranges of the species in n such genera; then there are $2n$ species. The pair of species forming each genus has a pair of ranges that are nested, lapped, or disjunct (see Figure 3.8). In what follows, pairs that are nested or lapped are grouped together and described simply as overlapping; that is, pairs are now classified dichotomously, as *overlapping* or *disjunct.* The object is to compare the proportion of overlapping pairs among congeneric species-pairs with the proportion among unrelated species-pairs, arbitrarily constructed by putting together two species picked at random from the whole set of $2n$ species.

Clearly, from the total of $2n$ species, it is possible to construct $n(2n - 1)$ different species-pairs by associating each species with every other. Of these pairs, n are the actual two-species genera with which we started; the remaining $2n(n - 1)$ pairs are arbitrarily formed, unrelated pairs. It is now straightforward to construct a 2×2 table showing a two-way classification of all the $n(2n - 1)$ possible pairs into, firstly, related (congeneric)

---

*The term is mathematical and the meaning is what it appears to be. We here disregard species with disjunct ranges, that is, ranges consisting of two or more separate subranges. The difficulty that sometimes arises of judging whether a range is disjunct or not is discussed in Chapter 9, page 279.

pairs versus unrelated pairs, and secondly, overlapping versus disjunct pairs.

As an example, the following 2×2 table shows results of this kind for the 37 two-species genera of Rhodophyta (red algae) growing on the Pacific coasts of North and South America (Pielou, 1978b).

	Species-pair	
	Congeneric	Unrelated
Overlapping	21	1175
Disjunct	16	1489

Under the null hypothesis that the two classifications are independent, the expected number of congeneric pairs that overlap is 16.4, which is less than the observed number. The deviation from expectation in this one table is not great enough to warrant rejection of the null hypothesis (χ^2 with continuity correction is 1.88). However, deviations in the same direction were found as well in the Phaeophyta (brown algae) and Chlorophyta (green algae) on the Pacific coast of the Americas, and also in all three algal phyla on the Atlantic coast. The probability that the differences between observation and expectation are all of the same sign in six tests, given that the null hypothesis is true, is 2^{-6}, or less than 2%. The weight of evidence therefore suggests that among littoral algae, congeneric species overlap each other more often than do unrelated species, implying that their geographic ranges are not affected by interspecific competition.

Simberloff (1970) approached the problem of interspecific competition on a geographic scale from an entirely different direction. For chosen taxonomic groups, he observed the mean number of species per genus, S/G, in each of many island biotas, with that of the mainland species pools from which these island biotas were presumably drawn. If the number of species per genus tends to be large, one probable inference is that congeneric species, because of similarity due to their common ancestry, have similar ecological requirements and similar dispersal capabilities, and hence tend to be found together. Another possibility is that the congeneric species have evolved autochthonously, as descendents of a common immigrant ancestor. If the number of species per genus tends to be small, the inference is that congeners tend not to occur together, probably because of interspecific competition.

Before it can be decided whether an observed S/G is "large" or "small," it is necessary to determine what its expected value, say $E(S/G)$, would be if the species on an island were merely a randomly drawn subset of the species on the nearby mainland. It is not correct to assume that $E(S/G)$ is the same as the mean number of species per genus in the mainland pool of species. In fact, $E(S/G)$ is less than this, as a simple example will show.

Imagine a mainland biota consisting of two genera, A and B, each with two species, so that for the mainland, $S/G = 2$. Let the species be called A_1, A_2, B_1, and B_2. Now suppose that two of these four species are picked at random to populate an island. There are six possible pairs, and under the null hypothesis they are equiprobable. They are A_1A_2, A_1B_1, A_1B_2, A_2B_1, A_2B_2, and B_1B_2. It is seen that the mean number of species per genus in these pairs is 8/6 or 1.33. Evidently, therefore, even if the species on an island are a wholly random subset from the mainland pool, the island biota will appear more diverse because each genus will tend to be represented by fewer species than on the mainland.

Simberloff estimated values of $E(S/G)$ for given islands and mainlands by computer simulation; (to guard against hidden biases, he also explored the effect on such estimates of different taxonomic treatments of the biotas, incomplete species lists for island and mainland, and incorrect identification of the region of the mainland serving as source area). He found that in most cases, for a given mainland pool, $E(S/G)$ increased approximately linearly with the size of the island biota drawn from the pool. He then compared observed and expected values of S/G for the biotas of 180 islands, each apparently stocked from one of 23 different mainland regions in various parts of the world. The taxonomic groups considered were chiefly land birds and vascular plants.

The observed S/G was found to exceed expectation in 70.6% of the 180 islands. Simulation tests had shown that under the null hypothesis the distribution of S/G was approximately symmetrical and hence that one would expect a roughly 50:50 split between islands whose S/G exceeded expectation and those that fell short of it. The probability of the observed result, given the null hypothesis, is less than 0.1%, and the evidence therefore very strongly suggests that competitive exclusion is not influencing the composition of these island biotas, at least so far as land birds and vascular plants are concerned. On the contrary, related species occur together unexpectedly often.

It will be interesting to find out whether Simberloff's (1970) and Pielou's (1978b) observations are the rule or the exception. So far as they go, they imply that competitive exclusion on a geographical scale does not take place. The best known case in which it (presumably) did occur resulted in the many extinctions of large grazing mammals in South America in the Pliocene, when the Isthmus of Panama formed and provided a migration route into South America for competitively superior invaders from the north (Simpson, 1950; and see Chapter 3, page 57). These extinctions are generally regarded as manifestations of competitive exclusion and there seems no reason to doubt the correctness of this view. Perhaps further work will show in what circumstances and among what kinds of organisms "geographic" competitive exclusion is likely to take place.

Interspecific Competition in Different Latitudes

We come now to the problem of whether interspecific competition* on an ecological scale proceeds with the same intensity in communities in different geographic regions, particularly at different latitudes. The reason for supposing that it does not is that the latitudinal gradient in the "diversity" (species richness) of nearly all kinds of communities is, according to one explanation, caused by the fact that the intensity of competition varies with latitude.

The argument has two versions. According to one, the presence of large numbers of weakly differentiated species in communities of the wet tropics must cause interspecific competition to be very intense. According to the other, the presence of numerous similar species implies that competition is weak (otherwise, how could they coexist?). Obviously the stalemate can only be resolved if some measure of the "average intensity" of competition can be found that does not depend on numbers of species.

One possibility is to examine communities on environmental gradients, along which species become sorted into zones. An easily visualized example is the altitudinally zoned vegetation of mountainsides. It is reasonable to suppose (and in at least one ecological context it has been experimentally demonstrated; see Connell, 1961) that zonation on a gradient is not caused only by the confinement of each species to a zone determined by its environmental tolerance limits, but also by competition with the species in contiguous zones. That is, a species may be constrained to occupy a narrower zone than its physiological tolerance limits would permit because of competition with species above and below it on the gradient. If this is so, one would expect zones to have abrupt boundaries where competition is intense and to blend into each other where it is mild. As a result, there would be a more rapid species turnover along

---

*Because of interminable debates among ecologists on what they mean by "competition," an explicit definition is necessary. I use the following: *Competition takes place when the growth of a population, or any part of it, is slowed because at least one necessary factor is in short supply* (Pielou, 1974, page 203). This definition embraces both *contest (interference)* and *scramble (exploitation)* (ibid, page 251).

the gradient (equivalently, a more rapid replacement of species) in regions where competition is intense.

The dependence of turnover rates on latitude has been investigated by Huey (1978) who studied the altitudinal zonation of species of lizards, snakes, and frogs on mountainsides from Costa Rica (at 10°N latitude) to California (at 40°N latitude). To arrive at an inverse measure of the species turnover rate, he computed the coefficient of community,* C, between faunas separated by each of several, chosen, vertical distances and then averaged the coefficients so obtained to get their mean, $\bar{C}$; it amounts to a measure of the mean faunal similarity to one another of d fferent altitudinal zones. He found clear evidence that $\bar{C}$ increases— equivalently, that species turnover rate decreases—with increasing latitude. The interpretation to be put on this result is not entirely clear however because Huey also found that the numbers of species in the three groups decreased with increasing latitude. Hence the decrease in turnover rate need not be solely due, although, as he shows, it is probably primarily due, to increasing latitude. So far as it goes, the evidence suggests that (among lizards, snakes, and frogs at any rate) interspecific competition is more intense in tropical than in temperate latitudes.

Unambiguous evidence on geographical variation in the abruptness of altitudinal zonation is hard to obtain, however, because of still another complicating factor. The abruptness of zone boundaries—equivalently, the narrowness of ecotones—is affected by the steepness of a slope as well as by interspecific competition. On an altitudinal gradient, the horizontal width of an ecotone is, of course, automatically narrower on a steep than on a gentle gradient. The vertical width of an ecotone is also likely to be affected by the steepness of the gradient, but in what way is difficult to judge. On a steep gradient an ecotone may be wider because of continual cross-invasions between zones only a short horizontal distance apart (Pielou, 1975). Alternatively, on a steep gradient it may be narrower, as suggested by empirical observations by Beals (1969); he examined plant zonation on altitudinal gradients in the semiarid highlands of Ethiopia and found that ecotones were narrower, in the vertical direction, on steep than on gentle slopes. To ascribe this result to more intense interspecific competition on steep slopes is not to explain why competition should be more intense on steep slopes, even if it is. More research is clearly needed on the facts of geographical variation in altitudinal zonation, and on the interpretation of these facts.

---

*The coefficient of community is defined as $C = 2s_{XY}/(s_X + s_Y)$ where s_X and s_Y are the numbers of species in zones X and Y respectively, and s_{XY} is the number common to both zones.

Another approach to the problem of the way in which interspecific competition among plants varies in intensity from one region to another has been made by Grime (1974, 1977). He notes that the total plant biomass in any habitat is limited by a combination of factors of three different kinds, one biotic and two abiotic.

The single biotic factor is competition, both within and between species.

The two classes of abiotic factors are (1) *stress* factors, which consist in a lack of conditions or substances necessary for plant growth, such as warmth, light, water, and nutrients; and (2) *disturbance* factors, events that damage or destroy plants, such as fire, grazing and browsing, wind damage, soil erosion, disease, and the like.

In any tract of vegetation, total biomass is governed by a combination of factors belonging to these three classes. Where one class of factors dominates, natural selection has led to the evolution, in the majority of the species, of characteristic adaptations. Thus selection for competitive ability tends to favor perennial plants (trees, shrubs, and herbs) with large, short-lived leaves that form a closed canopy. Plants selected for stress-tolerance include long-lived dwarf forms, often with small, long-lived, sclerophyllous or needlelike leaves. Plants selected to withstand frequent disturbance are the so-called "ruderal" plants (colloquially, "weeds"); typically, they are small in stature and short-lived, many being annuals; also, compared with plants in the other two classes, reproductive tissue forms a large proportion of their annual biomass production.

Grime (1974) applied his three-way classification to individual plants and to the plant biomass in small (meter square) quadrats. As a measure of a plant's stress tolerance he used an inverse measure of its growth rate, experimentally determined; as a measure of its competitive ability he used an index computed from various morphometric measurements. His classification can therefore be applied only to small tracts of vegetation whose species can be studied carefully and in detail. And its use entails accepting his assertion that certain morphological properties of a plant do indeed measure its competitive ability.

However, the classification could be applied crudely on a regional scale by simply translating Raunkiaer's life forms (see page 203) into Grime's classes as follows. We may, very roughly, equate Raunkiaer's phanerophytes with Grime's competitive plants, Raunkiaer's therophytes with Grime's disturbance-tolerant plants, and all the remaining life forms of Raunkiaer's classification with Grime's stress-tolerant plants. The correspondence is certainly not exact but the procedure has two advantages. First, use of three classes rather than five allows life-form spectra to be more compactly portrayed and more easily intercompared; a three-class

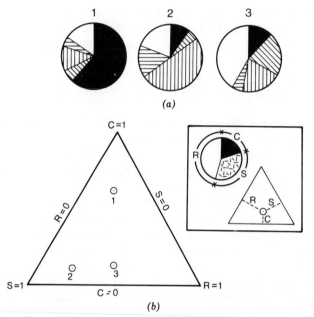

Figure 7.5 (*a*) Three life-form spectra represented as pie diagrams. The black and white sectors show, respectively, the proportions of phanerophytes and therophytes, and the three hatched sectors show the proportions of chamaephytes, hemicryptophytes, and geophytes in the floras of: 1, Seychelles Islands; 2, Central Switzerland; 3, the Libyan Desert. (Data and attributions are in Dansereau, 1957, page 70.) (*b*) Representation of these three floras as points, correspondingly numbered, in a single triangular coordinate frame. The inset figure shows how the proportions of the three recognized classes, as shown in the pie diagram, define a point in the triangle. The proportion of phanerophytes is C (for competitors), of therophytes is R (for ruderals), and of the remaining forms is S (for stress-tolerant plants). For every point in the triangle, $C + R + S$ = constant. (The form of presentation, and the symbols, are as in Grime, 1977.)

spectrum can be represented by a single point in a triangular coordinate frame, rather than by an elaborate pie diagram (see Figure 7.5). Second, it emphasizes the importance of interspecific competition in determining the morphology of plants and the structure of vegetation; probably most botanists would concede that phanerophytes tend to be the plants most tolerant of competition. Thus a triangular scatter diagram could form a useful summary of the range of vegetation types in a geographic region.

*4. MAPPING INFRASPECIFIC VARIATION

It is probably true to say that the majority of plant and animal species with extensive geographic ranges exhibit infraspecific variation. The variation may be phenotypic or genotypic and without definitive experiments it is usually impossible to decide which. Such terms as "race," "form," "variety," and "subspecies" are used to describe the variants. Usually, though not necessarily always, the different forms exhibit some degree of geographic segregation.

In some cases a unidirectional trend in a morphological character appears to match a geographic trend in an abiotic factor. Often, numerous unrelated species exhibit the same trend. Examples of this phenomenon have been summarized in several traditional biogeographic "rules." Examples are (see Ager, 1963, for details):

Allen's Rule. Individuals of many mammal species have smaller body extremities (for example, limbs, tail, ears) in cold than in hot climates; presumably this results from the selective advantage of a shape that conserves body warmth in the cold and dissipates it in the heat.

Bergmann's Rule. In warm-blooded animal species, individuals tend to be larger in cold than in hot climates. This, again, is probably related to the conservation of body heat (Guilday, Martin, and McCrady, 1964). The principle is believed to apply in reverse to poikilothermic animals.

Gloger's Rule. In many species of mammals, birds, and insects, individuals are dark colored in humid climates and comparatively light in dry climates. Presumably this is a camouflage adaptation; arid habitats provide light-toned backgrounds.

Jordan's Rule. In many fish species, for example, *Gadus morhua* (cod), the number of vertebrae is lower in individuals hatching in warm water than in cold.

Another, unnamed rule states that the body size of marine mollusks tends to increase with increasing salinity. This rule is based on observations on *Buccinum undatum, Littorina littorea, Mytilus edulis, Cardium edula, Mya arenaria,* and *Macoma baltica* in the North Sea and the Baltic.

These miscellaneous rules are only crude, general statements based (sometimes) on selected evidence. Detailed biogeographic studies require methods for mapping the geographic distributions of the infraspecific forms of a species. Numerous methods have been devised and they fall into two classes. Methods of the first class merely present data without judging it; the presentation is often by means of isophene (contour) maps; tests can be done separately, if desired, to judge the significance of observed geographic variation. Methods of the second class are designed to test the homogeneity of particular geographic subpopulations of a species as part of a classification and mapping procedure; the final map then shows the geographic ranges of apparently homogeneous infraspecific groups.

The following paragraphs describe one useful method of each kind from the multitude that have been proposed.

Trend Surface Mapping

Suppose some measurable morphological character of a species has been observed on a sample of individuals at each of a number of different sites. If the character varies geographically, the way in which it varies can often be most clearly displayed on an isophene (contour) map. Numerous algorithms exist for automatic contour mapping by computer, some so complicated that most users must accept them on trust. However, a straightforward way of constructing a trend surface map, as it is called, is to fit a polynomial surface. The surface, which may be linear, quadratic, cubic, or of still higher degree, is fitted to the observations so that the sum of squares of deviations of the surface from the observed means at the sampling sites is a minimum. Computer programs exist for achieving this.

Writing x and y for the geographic coordinates of a point in the area being mapped, and z_n for the height of the polynomial surface of degree n above the point, the equations of the various surfaces are as follows (all symbols apart from x, y, and the z_n are constants to be determined as part of the fitting):

For a linear surface (a plane):

$$z_1 = A + B_{10}x + B_{01}y;$$

For a quadratic surface:

$$z_2 = a + b_{10}x + b_{01}y + b_{20}x^2 + b_{11}xy + b_{02}y^2$$

or, more compactly:

$$z_2 = a + \sum_{i=0}^{1} b_{ij}x^iy^j + \sum_{i=0}^{2} b_{ik}x^iy^k$$

where $j = 1 - i$ and $k = 2 - i$.

For a cubic surface:

$$z_3 = \alpha + \sum_{i=0}^{1} \beta_{ij}x^iy^j + \sum_{i=0}^{2} \beta_{ik}x^iy^k + \sum_{i=0}^{3} \beta_{il}x^iy^l$$

where $j = 1 - i$, $k = 2 - i$, and $l = 3 - i$.

And so on.

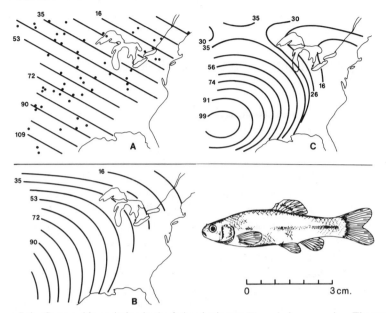

Figure 7.6 Geographic variation in the fathead minnow *Pimephales promelas*. The variate mapped is the completeness of lateral line, CLL. The isophenes (contours), which are labeled with percent CLL values, are those of fitted trend surfaces, linear in A, quadratic in B, cubic in C. The dots in map A show the 54 sample stations. (Redrawn from Marcus and Vandermeer, 1966.)

Isophene maps showing the contours of, respectively, linear, quadratic and cubic surfaces, all fitted to the same body of data, are shown in Figure 7.6. The data (from Marcus and Vandermeer, 1966) relate to a morphological character of *Pimephales promelas* (fathead minnow), namely the so-called CLL value, the percentage "completeness of the lateral line" (the percentage of scales in the fish's lateral line row that have pores). The data were obtained from 1087 individual minnows, caught at the 54 sites shown by dots in map A in the figure. An analysis of variance had shown that geographic (between-site) variation in the CLL values exceeded local (within-site) variation, and hence that the observed geographical variation is unlikely to be due merely to chance.

The fitted linear, quadratic, and cubic surfaces, shown in maps A, B, and C, accounted for 78%, 83%, and 90% respectively of the observed variation. These percentages are $100R^2$, where R is the multiple correlation coefficient between the height of the trend surface at a sample site and the observed variate values. Vandermeer (1966) fitted trend surfaces to other characters of the minnows besides the CLL, for instance, predorsal length, length and depth of caudal peduncle, length of eye, and angle of mouth. It is interesting to note that although these characters exhibit geographic variation, they do not vary concordantly.

Contiguity Partitioning of a Geographically Variable Species-Population

When a species exhibits geographic variation, it is always possible to define distinct classes such that every individual can be assigned to only one class. Members of the different classes may differ from each other in a qualitative character; for instance, human beings can be classified according to blood group. Or they may differ in some quantitative character, such as size, which varies continuously; then classes, for example size classes, have to be defined by choosing suitable, albeit arbitrary, class boundaries. A classification is often based on several characters some of which vary continuously while others are discrete. In any case, the process of assigning each of a number of conspecific individuals to a defined infraspecific category is familiar to all field biologists. In what follows, we call the recognized classes *morphs*. The differences among morphs may be discrete or continuous. Then a *geographically variable species is, by definition, one in which species-populations at different locations have different relative proportions of the morphs.*

We now wish to partition the area occupied by a species (or some part of the area) into regions in such a way that the population within each region is homogeneous. This is the same task as was tackled in Chapter 1 (see page 16) where the purpose was to subdivide a faunal (or floral) province into subprovinces on the basis of faunal (or floral) resemblances and differences. There, however, no objective way was suggested for ensuring that sample sites classed together because of their resemblance to each other were close together on the ground. Here it is shown how objectivity can be improved by considering only *contiguity partitions;* the term has its obvious, intuitive meaning and is defined precisely below. Subjectivity is not completely avoidable, however, since it is still necessary to choose how an acceptable contiguity partition is to be performed.

The following paragraphs discuss a procedure suggested by Gabriel and Sokal (1969) for carrying out the two steps required in a geographical-biological classification. There are two tasks: (1) to divide an area into geographically connected subareas*; and (2) to subdivide a variable population into internally homogeneous subpopulations. The two proc-

---

*In this account the word "subarea" denotes any geographically connected set of sites whose homogeneity is to be tested. The word "region" denotes a subarea that has passed all tests and is accepted as part of the final geographical-biological subdivision sought.

esses are conceptually separate but we require their results, so far as possible, to be congruent. That is, each geographically connected region is to contain the whole of one homogeneous subpopulation and no more.

Therefore the results of the two processes must be repeatedly compared with each other. In the description (in two numbered parts) below, the geographical partitioning of an area into connected subareas is dealt with first. In the account, mention is made of the need to test whether populations within subareas are homogeneous, but a digression on how the test is done is undesirable until the account is complete. Therefore the test is described subsequently.

We are given a map of the whole area concerned, showing the sampling sites from which data have been obtained. The datum from any one site is a list of the relative proportions of the different morphs found in a sample from that site.

We assume at the outset that the population has been tested (as described later) and found to be nonhomogeneous as a whole.

1. Partitioning the Area. The swarm of scattered points representing sampling sites on a map of the area should first be linked up by joining every *contiguous pair*. Two points, say *P* and *Q*, are contiguous if and only if no other point falls within a circle having the line *PQ* as a diameter (equivalently, *P* and *Q* are contiguous provided there is no point *R* such that angle *PRQ* is obtuse). In Figure 7.7 all contiguous pairs of points are

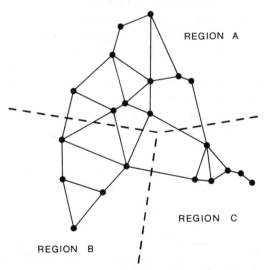

Figure 7.7 A Gabriel-connected graph divided into three regions. Further details are in the text.

joined by solid lines. The result may be called a Gabriel-connected graph.
Note that all points are linked to at least one other.

Any partition that separates the points into mutually exclusive subsets
each of which is a *connected set* is a *contiguity partition*. In a connected
set every point is linked to at least one other in the same set. In Figure 7.7,
the sets of points in regions A, B, and C are each connected sets.

It must now be emphasized that there is no unique, "natural" way of
performing a contiguity partition. The process only becomes objective
after the classifier has *chosen* the rules he will observe. And formal rules
should be used only after common-sense partitions have been made; thus
one would not ignore a mountain range cutting across the area to be
partitioned. To subdivide areas that appear uniform, a systematic approach should be adopted. Possible ways to proceed are suggested by
Gabriel and Sokal (1969). The aim is to construct as coarse a partitioning
as possible subject to the constraint that the set of sites within each
subarea is to be internally homogeneous. One may either begin with a
subjective partitioning that is very coarse and then subdivide those of its
areas that turn out, on test, to be nonhomogeneous; this is continued until
the coarsest acceptable partitioning is reached. Alternatively, one may
begin with a fine partitioning, in which all subareas are found (by test) to
be homogeneous, and then combine them in various ways until no further
combinations can be made without destroying the internal homogeneity of
at least one set.

One can also, if one chooses, adopt one or both of two rules that
increase the connectedness of each set, although this may result in some
sample sites being excluded from all the regions finally recognized. One
can require that α, the *index of connectivity,* of each region exceed some
chosen minimum. And one may require that λ, the *index of separateness*
of every region (from all others) exceed some chosen minimum. These
indices are defined as follows.

Let the number of sampling sites in a region be l.

Let the number of contiguity pairs wholly within a region be c.

Then α for the region is defined as

$$\alpha = \frac{c - l + 1}{2l - 5} \qquad \text{for } l = 3, 4, \ldots .$$

It is seen that the minimum value of α, $\alpha = 0$, is the index of connectivity
of a "string" of sites for which $c = l - 1$. Gabriel and Sokal (1969)
obtained no explicit formula for the maximum attainable value of α, but
found empirically that $\alpha_{\max} \approx \frac{2}{3}$ for $l < 3$.

To define the index of separateness, λ, for a given region, say R, we must first define a contiguity triangle; it is a set of three points of which each pair is contiguous. Now put t for the number of contiguity triangles that have exactly two points in R (and therefore the third outside R). Then

$$\lambda = 1 - \frac{t}{2c}.$$

It is easy to see that $0 \leqslant \lambda \leqslant 1$.

For the three regions in Figure 7.6, α and λ are found to be as follows.

	c	l	t	α	λ
Region A	14	10	1	0.33	0.96
Region B	6	5	1	0.40	0.92
Region C	7	6	0	0.29	1.00

To reiterate the salient parts of the foregoing discussion: The area to be subdivided is subjected to a contiguity partition based on common-sense considerations. The classifier then chooses an algorithm that permits the initial classification to be adjusted until the regions finally recognized are as large as possible subject to the constraint that the set of sites within each region must be homogeneous. The result is not unique; obviously it cannot be. The classifier also has the option of specifying minimum acceptable values for α and λ.

2. Testing for Within-Region Homogeneity. The datum for each site is the frequency distribution of the number of individuals in each recognized morph in a sample from that site. All the tests to be done are therefore straightforward tests of the homogeneity of observed frequency distributions. That is, if an area whose homogeneity is to be tested contains c sites, and there are r different morphs, one is testing the homogeneity of an ordinary $r \times c$ contingency table. In what follows, log likelihood ratio tests are used. The procedure is demonstrated below.

The important point to notice is that the homogeneity of subjectively chosen subsets of sites is to be tested. When statistical tests are applied to subjectively chosen (as opposed to random) data, spurious "significances" are inevitable unless precautions are taken. Moreover, many such tests are to be carried out, and the more there are, the greater will be the number of false rejections (due to chance) of the null hypothesis of homogeneity. We therefore require a procedure for carrying out several tests simultaneously on subjectively chosen partitionings of the sites. The method is as follows (Gabriel, 1966; Gabriel and Sokal, 1969).

First test the homogeneity of the initial, undivided set of data to judge whether any subdivision into regions is called for. The test is done by computing a test criterion, G (see below), that, under the null hypothesis, has a χ^2 distribution with $(r - 1)(c - 1)$ degrees of freedom (DF); (c here denotes the total number of sites before partitioning). Deciding the outcome of the test entails looking up a critical value, say ζ, of the test criterion in a table of percentage points of the χ^2 distribution. That is, we obtain from standard statistical tables the value of ζ such that

$$\Pr\{\zeta \geqslant \chi^2 \mid (r - 1)(c - 1)\} = \alpha$$

where $100\alpha\%$ is the chosen significance level (the probability of falsely rejecting the null hypothesis of homogeneity; equivalently, the probability of a type I error).

In all subsequent tests we are concerned with several subareas and test the homogeneity of all of them simultaneously.

Suppose the total area has been partitioned into m subareas. Then there will be m contingency tables to test. All have r rows; the jth has c_j columns and $\sum_{j=1}^{m} c_j = c$. Suppose the value of the test criterion for the jth table is found to be G_j. Then the null hypothesis, H_0, that all subareas are homogeneous, should be accepted if and only if

$$\sum_{j=1}^{m} G_j \equiv G_P < \zeta$$

where G_P denotes the test criterion for the simultaneous test.

Notice that the same critical value, ζ, is used for testing G_P as for G, the test criterion for the whole, unpartitioned set of sites. If the m separate contingency tables (one for each subarea) were independent—which they are not—then, given H_0, G_P would be a χ^2 variate with $(r - 1)(c - m)$ DF; that is, the DF would be the sum of the DF's of the separate tables and, since $(r - 1)(c - m) < (r - 1)(c - 1)$, the critical value would be less than ζ.

However, the m tables are not independent. On the contrary, they were obtained by subjectively partitioning a larger table. This is the reason one must use ζ as critical value. It ensures that the probability of a type I error is $100\alpha\%$ *at most*. The probability of a type II error (accepting H_0 mistakenly) is higher than it would be if the subareas were independent of each other, or equivalently, if the partitioning had been done blindly. But this is the price that must be paid for being allowed to inspect the data and sort the sites into groups before doing the tests. It should never be forgotten that "ordinary" statistical tests do not apply to consciously (as opposed to randomly) selected data.

Table 7.5 A 3 × 5 contingency table showing the observed frequencies of three morphs in samples from five sites. (The data are artificial. The definitions of $L_{i.}$ and $L_{.j}$ are given in the text.)

		#1	#2	#3	#4	#5	Totals	$L_{i.}$
				Sites				
	1	30	33	25	12	20	120	186.87
Morphs	2	60	62	48	18	31	219	334.37
	3	90	88	77	10	18	283	389.15
	Totals	180	183	150	40	69	622	910.39
	$L_{.j}$	182.05	188.06	150.83	42.68	73.76	637.38	

A numerical example will illustrate. See Table 7.5.

Write n_{ij} for the observed frequency in the (i, j)th cell of the table for $i = 1, \ldots, r$ and $j = 1, \ldots, c$. In the example, $r = 3$ and $c = 5$. It is necessary that all $n_{ij} \geq 5$. To test the homogeneity of the whole $r \times c$ table, calculate the log likelihood ratio, G, as follows.

Denote the ith row total by $n_{i.}$ and the jth column total by $n_{.j}$. The grand total is $n_{..}$.

For the ith row, compute

$$L_{i.} = n_{i.}\ln n_{i.} - \sum_{j=1}^{c} n_{ij} \ln n_{ij}.$$

For the jth column, compute

$$L_{.j} = n_{.j} \ln n_{.j} - \sum_{i=1}^{r} n_{ij} \ln n_{ij}.$$

(The values of $L_{i.}$ and $L_{.j}$ are shown in the table; also shown are $\sum_i L_{i.}$ and $\sum_j L_{.j}$.) Next, find

$$LL_c = n_{..} \ln n_{..} - \sum_{j=1}^{c} n_{.j} \ln n_{.j} = 921.92$$

and

$$LL_r = n_{..} \ln n_{..} - \sum_{i=1}^{r} n_{i.} \ln n_{i.} = 648.92.$$

Then the desired test criterion is

$$G = 2\left\{LL_c - \sum_{i=1}^{r} L_{i.}\right\} \qquad \text{or} \qquad G = 2\left\{LL_r - \sum_{j=1}^{c} L_{.j}\right\} = 23.06.$$

These two equations yield the same result and check each other. Equivalently,

$$G = 2\left\{n_{..} \ln n_{..} - \sum_i n_{i.} \ln n_{i.} - \sum_j n_{.j} \ln n_{.j} + \sum_i \sum_j n_{ij} \ln n_{ij}\right\}$$

but this is a less useful form if the table is to be partitioned.

Now find, from a table of percentage points of the χ^2 distribution with $(r - 1)(c - 1) = 8$ DF, the value of ζ for which

$$\Pr\{\zeta \geqslant \chi^2 \,|\, 8\} = 0.05 \qquad \text{(the chosen significance level).}$$

It is found that $\zeta = 15.5$. Since the computed test criterion $G = 23.06 > \zeta$, we reject the hypothesis that the five sites are homogeneous, that is, that the relative proportions of the three morphs are the same, apart from sampling errors, at all the sites.

Now consider two possible partitions, A and B, of the sites into two subsets. Denote the subsets by A1 and A2 in the first case, and by B1 and B2 in the second.

Partition A is:

$$\frac{\text{subset A1}}{\text{\#1 \quad \#2}} \quad \frac{\text{subset A2}}{\text{\#3 \quad \#4 \quad \#5}};$$

Partition B is:

$$\frac{\text{subset B1}}{\text{\#1 \quad \#2 \quad \#3}} \quad \frac{\text{subset B2}}{\text{\#4 \qquad \#5}}.$$

It is found that $G_{A1} = 0.17$ and $G_{A2} = 18.16$. Their sum, $G_A = 18.33$ and this exceeds $\zeta = 15.5$. Therefore partition A is unacceptable; it has not yielded homogeneous subsets.

However, $G_{B1} = 0.41$, $G_{B2} = 0.02$, and $G_B = 0.43$. Since the sum $G_B < 15.5$, the critical value for the simultaneous tests, partition B is acceptable; both subsets can be regarded as internally homogeneous.

To conclude this chapter, it cannot be overemphasized that the devising of spatial classification methods is a very active research field. Choosing among the methods is unavoidably subjective and the temptation to assume that the most recently invented is necessarily the best should be avoided.

Chapter Eight

DISPERSAL, DIFFUSION, AND SECULAR MIGRATION

Dispersal—using the word in a general sense—is, quite obviously, a topic of basic importance in all biogeographic investigations. Earlier chapters have mentioned it repeatedly, in other contexts, and this chapter examines the subject more closely. The aim is to illustrate the enormously varied processes involved, from those taking a few days to those taking tens of millions of years.

It is inconvenient to use the one word "dispersal" to describe such a variety of events, and Section 1 examines terminology and suggests a useful way of classifying the different kinds of dispersal. The two succeeding sections consider particular examples, on strongly contrasted time scales; thus Section 2 deals with the effects of continental drift on dispersal, and Section 3 considers rapid, long-distance "jump" dispersal.

1. THE THREE MODES OF SPREAD OF A SPECIES

We begin by defining the three terms in the title of this chapter. Usage is not standardized; however, the recognition of three different kinds of dispersal* (*sensu lato*) seems the best way of subdividing what would otherwise be too all-embracing a topic. For present purposes, therefore, the following definitions are used.

---

*It is important not to substitute the word "dispersion," even though "dispersal" and "dispersion" are synonyms in ordinary usage. "Dispersion" has been adopted as a (confusing) technical term by some ecologists to mean the spatial pattern of a population.

Jump-dispersal is the movement of individual organisms across great distances, followed by the successful establishment of a population of the original dispersers' descendants at the destination. The salient points are that the whole journey is completed in a short period of time, usually very much shorter than the life-span of an individual; and that the journey usually takes the dispersers across totally inhospitable terrain as, for example, when spiders are carried by air currents far across the open sea.

Diffusion is the gradual movement of populations across hospitable terrain for a period of many generations. Species that steadily expand their ranges can be said to be diffusing. Well-known examples are the American muskrat (*Ondatra zibethica*), expanding its range after introduction into central Europe; the starling (*Sturnus vulgaris*) and house sparrow (*Passer domesticus*), spreading westward across North America from introductions on the eastern seaboard; the fire ant (*Solenopsis saevissima*), spreading northward in the southern United States; and African "killer" bees (a race of *Apis mellifera*), spreading northward in South America following introduction into Brazil.

Secular migration (a term due to Mason, 1954) is diffusion taking place so slowly that the diffusing species undergoes appreciable evolutionary change in the process. The range of a secularly migrating species expands or shifts during very long, even "geological," time intervals. The species migrates into new environments and everywhere, at the same time, environments themselves are undergoing continuous secular change. Natural selection acting on the migrants therefore causes the descendent population in a new region to differ from its ancestral population in the source region. For instance, South American members of the Camelidae (the camel family) such as the llama (*Llama peruana*) and the vicuna (*Vicugna vicugna*) are descended from now extinct North American ancestors that made a secular migration, in the Pliocene, over the newly formed Isthmus of Panama.

It should be noticed that these three processes, which we may call, collectively, "spread," all result in changes in the geographic range of a species. A fourth mechanism, besides these three, also affects range, and that is local extinction. It is instructive to visualize the temporally changing range of a species. Obviously the patch (or patches) that represent its range on a map can expand, contract, or "creep." Expansion can result from all three modes of spread; contraction results from local extinction; and creep from a combination of spread and local extinction.

It should also be noticed that jump dispersal, diffusion, and secular migration, as defined above, describe three pure, extreme modes of spread. In practice, spread often consists of an ill-defined mixture of modes. Thus diffusion can sometimes be equated with a series of short jump dispersals; slow diffusion and rapid secular migration merge into each other.

As the terms have been defined above, "pure" diffusion consists in the crossing of totally hospitable terrain in a series of short steps, one per generation for many generations, and jump dispersal in the crossing of totally *in*hospitable terrain in a single jump. Clearly, however, the terrain to be covered by a spreading species can have any degree of hospitableness between these extremes. This brings us to the important topic of biogeographic *barriers*.

Barriers to spread can be of many kinds and of any degree of impassability. The most obvious barrier to terrestrial organisms is a body of water, especially a stretch of sea that cannot be circumvented. The question of how oceanic islands have been populated is, and has long been, a flourishing branch of biogeographic research and is briefly mentioned in Section 3. The most obvious barrier to marine organisms is land. We have already discussed (page 130) the way in which Beringia functioned alternately as a bridge for terrestrial animals and a barrier to marine ones during periods of marine regression, and vice versa during marine transgressions. Similarly the gap between North and South America was, before the isthmus formed, a route between the oceans for marine organisms and a barrier for terrestrial ones (see pages 32 and 151). Moreover the deep sea is a barrier to the shallow water benthos of the continental shelves (page 170).

Less obvious barriers can be equally difficult to cross. The alternate southward and northward shift of climatic zones that accompanied the glacials and interglacials of the Pleistocene ice age rendered land routes alternately usable and unusable by different groups of organisms. Thus the Bering land bridge was a corridor for ungulates only in periods when the climate was mild enough for appropriate fodder to be available (see page 131). Likewise in low latitudes: a tract of land may serve as a corridor for one group of animals during arid periods when it is savannah covered, and for another group during moist periods when it is forested.

Climatic change does more than create and destroy land routes. Marine climates have an effect on sea routes and hence on the spread of marine organisms. For example, according to Fuchs (1960), when the Antarctic ice sheet was larger than it is now, the cold-water zone surrounding it formed a barrier to the movement of warm-water species around the southern tip of South America.

Climate also affects the spread of plants and animals indirectly, through its effect on vegetation. It was remarked above that animals can only diffuse through a region where the appropriate food can be found. Vegetation also influences the spread of plants, especially wind-disseminated ones; their disseminules may be stopped by forests even though they can spread rapidly across open country.

It should be observed that in this discussion of barriers, no distinction has been made among barriers to jump dispersal, to diffusion, and to secular migration. A desert may be an impassable barrier to the slow diffusion of a mesophytic plant, but offer no resistance to the rapid passage of wind-blown seeds across it. The same species of plant might diffuse through, though it could not jump across, a dense forest. In the light of these considerations it is interesting that both Darlington (1957) and Lindroth (1963b), after weighing much evidence, concluded that, in general, plants spread more readily than animals. As a consequence, the modern distributions of plant species tend to reflect current conditions of climate and soil, whereas those of animals tend more to reflect the geographic and geologic history of a region.

This conclusion emphasizes how difficult it is, in attempting to explain the present-day geographic ranges of species, to know how much weight to attach to different factors; it is particularly hard to judge the relative importance of ancient and modern causes, and the extent to which the effects of events and conditions in the distant past have been obliterated by more recent events. The following sections, which consider the effects of processes occurring at very different rates, emphasize this.

2. THE EFFECTS OF CONTINENTAL DRIFT ON SPECIES SPREAD

Continental drift affects all three modes of species spread. Of itself, it accomplishes (for terrestrial organisms) the slowest form of secular migration, by gradually rafting species—and, indeed, whole communities—across enormous distances. Organisms whose observed distributions are directly attributable to this rafting are apt to be those for which long-distance jump dispersal is so remotely improbable as to be effectively impossible. The evidence of exceedingly slow drift has not, therefore, been obscured.

Several examples are described in Chapter 2. Among the animals mentioned were large, flightless vertebrates (for instance, *Lystrosaurus*, *Sphenodon*, living and extinct ratites, living and extinct marsupials) and a few invertebrates (for instance, planarians, soil nematodes) that cannot survive either drying or exposure to salt water. Many more examples could be quoted. Darlington (1957) has emphasized especially the extreme unlikelihood of freshwater fishes successfully dispersing across even narrow stretches of sea, and his arguments apply to every organism that must be in contact with a fresh, moist substrate—water or damp soil—at all times.

Typical of these are terricolous oligochaetes (earthworms). As Omodeo (1963) has shown, their present-day distributions are without doubt the direct and obvious outcome of continental drift. There are several Gondwana genera: for instance, *Microscolex* and *Eodrilus* occur in southern South America, South Africa, and Australasia; farther north, *Neogaster, Dicogaster, Nematogenia,* and *Wegeneriella* are common to west Africa and Central America. Farther north still, in what was Laurasia, the majority of earthworm species are the same on both sides of the Atlantic. Earthworms are, of all animals perhaps, the least likely candidates for transoceanic jump dispersal. They cannot survive submersion in salt water and neither can their eggs; they would be exceeding unlikely to survive as passengers on logs or floating mats of vegetation, and even if they did, they would have little chance of reaching an endurable habitat from the beach where their sea crossing would terminate. Since they live in the soil rather than on it, neither they nor their eggs are at all likely to become attached to birds' feet, and if they did they would quickly die of dehydration.

Plants whose present-day distributions still reflect Paleozoic geography all have seeds that seem unsuited to jump dispersal. Two Gondwana genera, *Podocarpus* and *Nothofagus,* are good examples (see page 47). Preest (1963) has investigated the dispersal characteristics of these genera and considers that, in both, jump dispersal occurs only occasionally and then not across the sea*; it happens when seeds are waterborne by flooding freshwater rivers.

Podocarpus, whose seeds are heavy, is chiefly dispersed by birds; depending on the species, the seed is enclosed in, or attached to, a succulent receptacle or peduncle that attracts frugivorous birds. Except on migration, terrestrial birds seldom travel far, and very rarely indeed across salt water; moreover a single seed would not suffice to bring about colonization of new territory since all *Podocarpus* species are dioecious.

The seeds of *Nothofagus* (southern beech), like those of *Podocarpus,* are not adapted for jump dispersal. Species of the genus have probably spread overland by slow diffusion, and have been stopped by the sea. The fact that the genus is confined to the Gondwana continents accords with this supposition. Fossil *Nothofagus* pollen, of Oligocene age, has been found at two sites in Antarctica. The modern disjunct range of the genus (see Figure 8.1) has resulted from the breakup of Gondwana.

Strong additional evidence (if it were needed) that *Nothofagus* has not

---

*For a dissenting opinion, see Carlquist (1974).

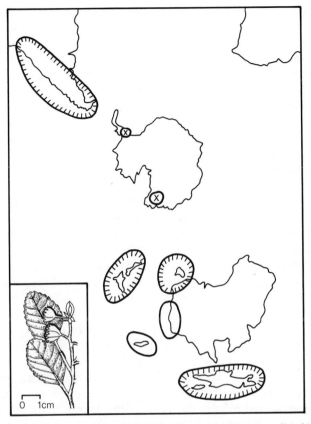

Figure 8.1 The modern distribution of the genus *Nothofagus* (southern beech). It occurs in the six outlined regions. In the four regions with hatched outlines, the aphid *Sensoriaphis* occurs with it. *Nothofagus* pollen, of Oligocene age, has been found at the two sites in Antarctica marked ⊕. (Adapted from Schlinger, 1974.) Inset: *Nothofagus truncata*. (Redrawn from Poole, 1950.)

spread by jump dispersal is due to Schlinger (1974) who found that closely related aphids, of a single genus (*Sensoriaphis*), infest *Nothofagus* in four of its subranges: South America, New Zealand, Victoria with Tasmania, and New Guinea. The modern aphid populations and their modern host trees are presumably lineal descendents of ancestral aphids and ancestral trees of late Cretaceous age (or earlier); they have, together, been passively rafted for the past 80 m.y. (that is, since New Zealand separated from Australia) with, perhaps, hardly any change of range from other causes. It is customary to suppose that light, buoyant, weak-flying insects like aphids are dispersed over long distances by air currents, and no doubt in many cases this is true. But it does not rule out the possibility that, for some species, the spreading process is the slow secular migration caused by the drifting continents. This is especially likely to be true of host-specific parasites whose spread must match that of their hosts. It is also likely to be true of the superparasites that attack the parasites, and in this example it apparently is. *Sensoriaphis* (as well as another, related podocarp aphid, *Neuquenaphis,* that occurs only in South America) is attacked, in both South America and Australia, by very closely related parasitoids belonging to the same primitive subfamily (Ephedrinae) of the Braconidae (order Hymenoptera).

Continental drift has affected the spread of plants and animals in various indirect ways, as well as directly, by rafting. Thus the separation and joining of landmasses have, respectively, broken and created the land routes (while creating and breaking the sea routes) used by secular migrants and diffusers. The opening and closing of the Bering routes, and of the Panama routes, have already been referred to frequently. The latitudinal displacement of landmasses has also affected migration and diffusion routes by shifting them into different climatic zones. And the uplifting of mountain chains, believed to occur when drifting tectonic plates collide, has provided long-distance diffusion-plus-dispersal routes across the latitudinal climatic zones. This last point has been noted by Raven and Axelrod (1972). They consider that the mountains of Malaysia and New Guinea, which were uplifted in Pliocene-Pleistocene time, provided a cool, high-altitude route for the diffusion of northern hemisphere plants into Australia. Among migrants they mention are the familiar northern genera *Veronica* and *Euphrasia* (both of the family Scrophulariaceae).

For organisms capable of jump dispersal, the influence of continental drift on diffusion cannot be clearly separated from its influence on jump dispersal. As landmasses drift toward each other, the number of dispersers that can jump the progressively narrowing gap steadily increases. Not until contact is made, of course, does a diffusion route for nondispersers come into being. Thus in the Central American region, as the two Americas approached each other, plants spread northward far more rapidly than did vertebrate animals until the land link closed. On balance, the spread of plants has been more pronounced from south to north than from north to south.

Likewise in the Indonesian region. The Australian tectonic plate drifted northward after becoming detached from Antarctica, and made contact with the Asian plate in the Miocene. Wallace's line is the line of contact (Raven and Axelrod, 1972). As over-water distances have decreased, the jump dispersal of plants between the two regions has been becoming steadily easier. But the land surfaces have not yet joined and the saltwater straits separating them are still a barrier to vertebrates. Therefore Wallace's line or, more accurately, Wallacea (see page 14), is much more of a zoogeographic than a phytogeographic barrier. Among plant genera that have crossed from the Australasian into the Oriental region, two well-known ones are *Eucalyptus* and *Casuarina*; though predominantly Australian, they occur also in Malaysia and in islands west of Wallacea (Good, 1974). The spread of species in the Far East has been chiefly north to south.

It is not clear why the spread of plants should be predominantly northward between the Americas and predominantly southward between Asia and Australia–New Guinea. It may merely be that South America and Asia were the richer source regions. It should be recalled (see page 133) that in Beringia the spread of terrestrial biotas has been chiefly from west to east, and of marine biotas chiefly from the Pacific to the Atlantic; and that the most likely explanation of these directional biases is simply that the numbers of species in the joined regions were unequal initially.

As landmasses drift away from each other, jump dispersal between them becomes steadily less frequent, of course. Evidence of the nonoccurrence of spread is harder to perceive than evidence of its occurrence. The clearest indication is obtained when sequences of fossil biotas are found that become steadily more different from each other the more recent they are. For example, the floras of South America and Africa have diverged as the gap between them has widened. Pollen floras of late Cretaceous age from Gabon (in west Africa) and from Brazil appear to be identical (Raven and Axelrod, 1974). Now, the floras of the two continents are entirely different.

Next consider the North Atlantic and recall first (see pages 29 and 32; also Cox, 1974 and McKenna, 1975) that the modern configuration of Holarctica differs strikingly from earlier configurations. The splitting of the early Tertiary "Euramerica" by the protoAtlantic ocean probably began about 49 m.y. ago (see Table 2.1, page 31). In the words of McKenna (1975), North America shifted its allegiance from Europe to Asia.

The biogeographic effects of the widening of the North Atlantic are particularly hard to interpret, for several reasons. The northern end of the Atlantic is surrounded by islands that might, for some species, serve as well-spaced stepping stones; thus Greenland, Iceland, and the Faeroes are at approximately equal intervals on a route from Labrador to Scotland, and north and east of Iceland lie Jan Mayen and, at a very high latitude, Svalbard. There may have been, however, in the early and middle Tertiary when the ocean itself was narrower (see Figure 8.2), more extensive tracts of dry land around the head of it that have since found-

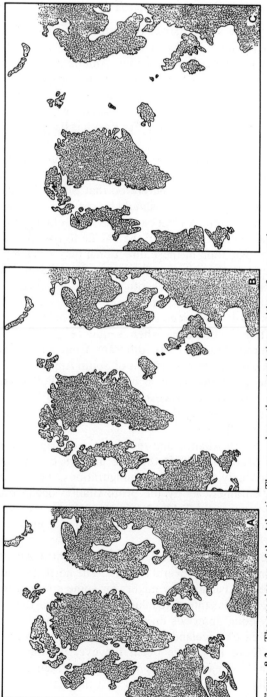

Figure 8.2 The opening up of the Atlantic. The map shows the past relative position of *present-day* landmasses; coastlines were not the same in the past because of changes in ocean floor topography and in the level of the sea, and at times there may have been much more land than now, forming landbridges. (A) Early Cretaceous. (B) Mid-Tertiary; volcanic lavas have formed Iceland and the Faeroes. (C) Late Tertiary; Jan Mayen has appeared northeast of Iceland as a result of volcanic action in the Miocene. (Adapted from Harland, 1969.)

ered (Einarsson, 1963). There is great disagreement as to whether biotic spread by diffusion was possible along some or all of the trans-Atlantic route at various times, and also over the widths, at different times, of water gaps that terrestrial organisms could cross only by jump dispersal.

Concurrently with these changes in the extent and spacing of the land areas, there were changes in their altitudes; the mountains of Greenland, Iceland, northern Britain, and Scandinavia were uplifted in the late Tertiary (Einarsson, 1963), and there were consequent changes in climate. Simultaneously, the climate of the whole world was cooling, a process that culminated in the Pleistocene glaciations. These may have destroyed all life in Greenland, Iceland, and the smaller islands; alternatively, a few hardy species may have survived in refugia (see Chapter 4, page 117). There is much disagreement on this point.

Thus there is great uncertainty about the biogeographic history of the lands around the North Atlantic. In northern latitudes there are no taxa analogous to, say, *Nothofagus* and its aphids at the southern end of the world, whose modern distributions can be almost wholly explained by continental drift. Presumably the populations of many northern taxa were split in two, and the parts rafted away from each other, when Laurasia broke up. But such populations have not been able to maintain their on-the-ground locations as the Atlantic widened; they have been forced to make alternate southward and northward migrations in response to climatic shifts and the advances and retreats of the ice sheets, and many have become extinct.

Interpreting fossil and modern biogeographic data from the North Atlantic region is therefore fraught with difficulty. Because of the uncertainty surrounding the geological history of the region, one might attempt to base a reconstruction of its physical past on a knowledge of current biogeography if only more were known about the rates, and chances of success, of the different modes of biotic spread under given conditions. But to argue from the biotic to the abiotic requires a much firmer knowledge of biotic spread than now exists, as well as of evolutionary rates in different taxonomic groups.

Attempts have been made. The distributions of freshwater fishes and freshwater bivalves have been taken to imply that North America and northwestern Europe were linked by a complete land bridge at some time in the Tertiary (references in Briggs, 1974). But according to Simpson (1947) and Darlington (1957), mammals moving between the Palearctic and Nearctic regions have used the Bering route, not the North Atlantic route, at least since the Eocene. The spread of plants, presumably by jump dispersal, continued via the North Atlantic route after the route had become impassable for mammals. As the climate cooled, the route was used by plants adapted to progressively cooler climates (Raven and Axel-rod, 1974).

There are different opinions as to which of the several modern water gaps between North America and northwestern Europe has constituted the most important barrier to biotic spread. Löve (1958), on botanical evidence, considered the Greenland-Iceland gap (Denmark Strait) to have been the most difficult to cross, and that Iceland was linked to Europe by a land bridge during the Riss glaciation.

Lindroth (1960, 1963a) regards Davis Strait (between Greenland and Baffin Island) as a far more important biogeographic barrier; and he regards Greenland, in spite of its proximity to North America, as part of the Palearctic region (contrast this with Figure 1.1). The beetle fauna of Baffin Island is closer to that of Siberia than to that of Greenland. Lin-droth hypothesizes that a land bridge linked Scotland, the Faeroes, Ice-land, and Greenland, perhaps intermittently, at least until one of the Pleistocene interglacials, and permitted the diffusion and sharing of biotas adapted to a somewhat milder climate than that of the present day. The land bridge, now submerged, forms the present Wyville Thomson sub-marine ridge. Having regard to supposed evolutionary rates (many fossil insects from the Eocene are different from their presumed descendants, but all those known from the Pleistocene appear to be identical with extant species), the absence of endemics suggests that the islands were stripped of their indigenous biotas by one of the early glaciations and were subsequently colonized, during an interglacial, via the land bridge. This land bridge did not cross Davis Strait. The colonizers are then assumed to have survived at least the last glaciation in situ. This theory is proposed by Lindroth to account for the otherwise inexplicable biotic contrast between Greenland and Baffin Island.

The effect of Davis Strait on the spread of plants can be summarized in the 2 × 2 table shown below, which condenses more detailed information in Lindroth (1960). Classifying plants first into those that have and those that have not crossed Davis strait, and second into those that are and those that are not easily dispersed over long distances, gives the following table:

| | | Adapted for jump dispersal | |
		Yes	No
Found on both sides of Davis Strait	Yes	26	34
	No	11	40

This constitutes significant evidence ($\chi^2 = 4.94$, $P \simeq 0.025$) that the two classifications are not independent; plants adapted for jump dispersal are more likely to have crossed the strait than those not so adapted. Even so, 34 "poor dispersers" have made the crossing.

In a later paper, Lindroth (1963a) explored the effect of Davis Strait on the fauna of Greenland and concluded that the strait is the "most effective faunal barrier of the entire circumpolar area." He compiled data for distributional spectra (see page 222) of five representative animal groups in Greenland: Macrolepidoptera (moths and butterflies), Araneae (spiders), and Aves (birds), all of which are flying or wind-dispersed animals; and Coleoptera (beetles, most of the Greenland ones flightless) and Collembola (springtails), which are ground dwellers. As may be seen in Figure 8.3, a large proportion of each group are Holarctic and may well have spread via Beringia. Of the non-Holarctic species and subspecies, the ground dwellers are chiefly Palearctic, and only among taxa capable of jump dispersal do Nearctic taxa predominate. This is further evidence of the importance of Davis Strait as a biogeographic barrier.

Many biogeographers would disagree with Lindroth that many (or, indeed, any) organisms could have survived the last glaciation (the Wisconsin-Würm) in Greenland, Iceland, and the Faeroes. For example, Savile (1956) considers that, except possibly for a few high arctic species that may have survived in unglaciated Peary Land (northernmost Greenland), all present-day Greenland plants must have invaded in the Holocene, across a water gap at least 300 km wide. Indeed, Lindroth's (1960) own data show that of 111 Greenland plants, 87 are Nearctic and only 24 Palearctic in origin.

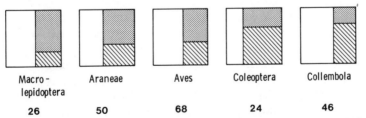

Macro-lepidoptera	Araneae	Aves	Coleoptera	Collembola
26	50	68	24	46

Figure 8.3 The subdivision into Holarctic (unshaded), Palearctic (hatched), and Nearctic (stippled) of the taxa (species and subspecies) in five representative groups of the fauna of Greenland. The number of taxa in each group is shown below its name. These are distributional spectra (compare Figure 7.4), but instead of being shown as pie diagrams they are represented in a way that emphasizes the division of the faunas first into Holarctic and non-Holarctic taxa (left and right), and then the division of the latter into Nearctic and Palearctic taxa (above and below). Notice that the ratio of Palearctic to Nearctic taxa is high among ground dwellers and low among active and passive fliers. (Data from Lindroth, 1963a.)

Böcher (1963) has shown that American plant species outnumber European ones in all land areas of Greenland except the southeast. He argues
that the present locations of American and European plant species reflect
the environments to which they became adapted in their respective home
continents. Immigrants from Europe required high humidity and a deep
winter snow cover, such as occur in the Alps, Norway, and Scotland;
these plants are restricted to southern Greenland and cannot succeed in
the dry, "continental" environment of northern Greenland. The colonists
that have established themselves in northern Greenland tend to come
from the relatively dry tundra of the Canadian Eastern Arctic.

The ranges of many Greenland plants, which are, of course, entirely
restricted to the icefree coastal strip, exhibit one of two contrasted patterns; on a map, they may appear U-shaped (occurring on the west, south,
and east coasts) or ∩-shaped (on the west, north, and east coasts). Doubly
classifying 75 species, first on the basis of their continent of origin, and
second on the shape of their range, gives approximately the following 2 × 2
table [data adapted from Böcher (1963) by excluding all species whose
ranges are not U-shaped or ∩-shaped]:

	U-shaped Range	∩-shaped Range
European origin	36	0
American origin	27	12

Performing a χ^2 test gives $\chi^2 = 11.00$, $P \simeq 0.0009$, which strongly supports
Böcher's thesis. Two typical range maps are shown in Figure 8.4; they
relate to *Minuartia Rossii*, a plant of the Canadian Arctic, which has
invaded Peary Land and has also reached northeastern Greenland and
Svalbard; and *Saxifraga stellaris*, a European species, which occurs in
Greenland only in the south, with a typical U-shaped distribution.

The foregoing discussion of North Atlantic biogeography should demonstrate how often the "facts" of geological and climatic history, biotic
dispersal, and evolution lead different workers to different conclusions,
none being wholly in accord with the data since the data themselves seem
irreconcilable. The whole subject still offers a tempting challenge.

3. DISPERSAL BY LONG-DISTANCE JUMPS

Earlier paragraphs have stressed the contrasting abilities to spread far and
rapidly of organisms that can, and that cannot, make jump dispersals over

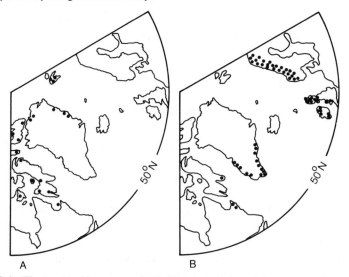

Figure 8.4 The geographic ranges of (*A*) *Minuartia* (*Arenaria*) *Rossii,* a plant of arctic North America whose eastward spread has been through the high arctic into northern Greenland and Svalbard; and (*B*) *Saxifraga stellaris,* an Eurasian plant whose westward spread has been via southern Greenland and thence to Baffin Island and Labrador. (Adapted from Porsild, 1957, and Hultén, 1958.)

long distances. In this section we consider jump dispersal, its biogeographic importance, and the evidence concerning it.

Composition of Waif Biotas

In the first place, it is obvious that oceanic islands, those that have never been linked by land to a continent, can only have acquired their biotas by jump dispersal. It probably accounts also for most or all of the biotas of once-glaciated islands, even if in pre-Quaternary time they had been joined to continents. A proportion of the biotas of many other regions may also be descended from immigrants that arrived by jump dispersal, but in every case it is debatable how large this proportion is. The kinds of regions that may owe an appreciable proportion of their biotas to jump immigration include continental fragments that drifted away from their parent continents a long time ago (for example, Madagascar and New Zealand); continental land areas that were laid bare by melting ice sheets; and mountain summits. As biotic environments, high mountains are just as much "islands" as are islands in the sea; the fact that two separate mountains in the same continent may have strikingly different biotas although they are linked by a "land bridge," demonstrates that land bridges do not automatically serve as highways for all kinds of organisms.

Biotas made up of long-distance immigrants are known as *waif* biotas and their diagnostic character is that they are *disharmonic*. That is to say, compared with biotas of continental areas, they contain a disproportionately high number of organisms adapted for jump dispersal.

The general characteristics of Pacific island faunas are a good illustration (Gressitt and Yoshimoto, 1963). They contain hardly any mammals except bats and rats; bats can fly, and rats have almost certainly been introduced by ships. There are many land birds, often of very restricted range, of which only a few are migratory. Among reptiles, skinks and geckoes are widespread; they are probably transported on flotsam; snakes are found only on islands close to continents. Amphibians and freshwater fishes are almost entirely absent. The dominant animals are insects; land mollusks, too, are well represented.

Probably most insects are brought in by air currents. Gressitt and Yoshimoto (1963), who took a series of samples of the aerial plankton over the Pacific, found a fairly close correspondence between the contents of their traps* and the insect fauna of the Hawaiian Islands. The main discrepancy was that Macrolepidoptera, Carabidae (ground beetles) and Curculionidae (weevils) were much scarcer in the aerial traps than on the islands. Probably, therefore, the majority of these heavier insects were brought to the islands as passengers, larval or adult, inside fragments of vegetation—twigs, leaves, seeds, bits of bark, and the like— carried by storm winds. Strong-flying insects, such as large moths and butterflies, and dragonflies, presumably immigrated under their own power, aided by wind.

Research on the manner in which organisms of all kinds have disseminated to oceanic islands, and how they have fared when they arrived there, has been going on actively since the time of Darwin and Wallace and before. The subject has been reviewed by Carlquist (1974). An illuminating way of looking at the vast (though very incomplete) array of available data is to consider in what way, and why, the characteristics of the biotas of different islands differ from each other. Often they differ markedly, even among islands whose climates and habitats are very similar.

*Among 1034 insects, spiders, and mites (excluding marine Heteroptera), caught in the aerial traps, the percentages of the different orders were as follows: Diptera, 40%; Hymenoptera, 18%; Homoptera, 17%; Coleoptera, 7%; Heteroptera, 4%; Araneida, 3%; Thysanoptera, 3%; Lepidoptera, 2%; Psocoptera, 2%; Neuroptera, 1%; and less than 1% each of Isoptera, Collembola, Acarina, Blattaria, and Ephemeroptera (Gressitt and Yoshimoto, 1963).

When due allowance has been made for the differences among islands in the mix of environments they provide for immigrants, two factors emerge as outstandingly important in determining their biotas. These are the length of time an island has been in existence, and its distance from a continent.

Consider first the effect of an island's age. Obviously, the older it is, the more time there will have been for autochthonous evolution of endemics (see Chapter 6, page 192). Oceanic islands vary greatly in age. Most tropical Pacific islands for which data are available are not older than early Eocene. Islands whose surfaces were above sea level before that time have now mostly disappeared as terrestrial habitats. They persist as flat-topped submarine sea mounts (guyots) that have been leveled by erosion and finally submerged (see Figure 6.10). Moreover it is likely that, in the southwest Pacific, the areal extent of dry land and the distances separating island groups from each other and from a continental landmass have not been constant (Whitmore, 1973). If this is so, the effects on an island's biota of age and distance from a continent may be hard to disentangle.

The same is true of the Atlantic islands (Wilson, 1963b, 1965). The youngest are those nearest the mid-Atlantic ridge (see Figure 8.5); they were formed by comparatively recent volcanic action along the ridge, when the ocean was already wide, and they are still the sites of active volcanoes. Successively older islands are nearer to one or other of the landmasses east and west of the ocean. They were formed when the Atlantic was narrower than it is now, and ocean-flooor spreading has rafted them away from the mid-ocean ridge, together with the continents, on the surfaces of their respective tectonic plates. Thus the youngest islands are also the most remote.

An island's biotic age can be considerably younger than its geological age; if it is denuded of life by a volcanic eruption, as Krakatoa was in 1883, the development of a biota has to begin anew. The youngest of the world's islands at present is Surtsey off Iceland, which first appeared in 1963.

The rate and nature of evolutionary diversification on islands of different ages is a topic whose existence can scarcely be more than acknowledged in a general book on biogeography. But two points particularly relevant to evolution on islands, and having to do with jump dispersal, deserve mention. The first of these is the evolution on islands of organisms with poor dispersibility from ancestors that could, and did, invade by means of long-range jump dispersal. (The phenomenon is not, in fact, limited to islands, but is well displayed on them.) Carlquist (1974) has investigated the phenomenon and offers much evidence of it, in both

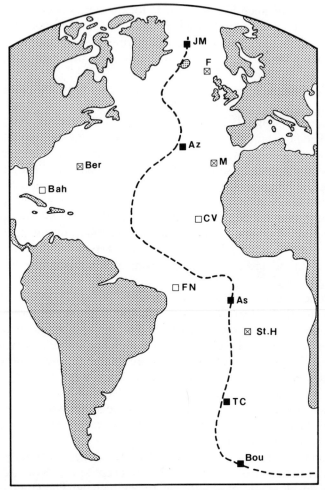

Figure 8.5   The arrangement with respect to the mid-Atlantic ridge (broken line) of islands of different ages. ■ Young islands (less than 20 m.y. old) with active volcanoes; JM, Jan Mayen; Az, Azores; As, Ascension; TC, Tristan da Cunha; Bou, Bouvet. ⊠ Islands with early Tertiary rocks; F, Faeroes; M, Madeira; Ber, Bermuda; St.H, St. Helena. □ Islands with Cretaceous rocks; Bah, Bahama; CV, Cape Verde; FN, Fernando de Noronha. (Adapted from Wilson, 1963b, 1965.)

plants and animals, as well as a choice of theories that might explain it. Presumably good dispersibility is either selectively disadvantageous on small islands because it leads to the wasteful loss of individuals (in plants, a loss of seeds); or else, good dispersibility is selectively neutral and tends to be lost coincidentally, as other characteristics are favored by natural selection.

The second point is that the species whose populations in each of a group of islands diverge most rapidly and most strongly from their mainland ancestors and from each other must be those for which dispersal "jumps" are least frequent. Rapid differentiation can only take place in an isolated population, one that is immune to frequent injections of ancestral genes. The degree of difference, from island to island in a group of islands, between related lineages varies enormously from one taxonomic group to another, and no doubt reflects the dispersibility of species in the different groups and hence the probable lengths of their periods of isolation. It has been found that upland species, especially mountain ones, tend to be the most sharply differentiated, and strand species the least; among plants, however, even strand species, which are probably the best jump dispersers, exhibit an appreciable amount of geographic variation (Fosberg, 1963).

Next, consider the effect of distance from the mainland on an island's biota. Since jump dispersers are very unequal in dispersal ability, one would expect the biotas of islands at great distances from any continental landmass to contain a disproportionately large number of the best dispersers; in islands closer to a mainland coast, there would be relatively more dispersers of comparatively limited dispersibility. Distance can thus be expected to have a sorting effect.

However, so many factors are at work in determining an island's biota that clear patterns seldom emerge; one can do little more than search for, and almost certainly find, support for one's pet theory in data whose almost every item is idiosyncratic. The lack of pattern is demonstrated, so far as angiosperms are concerned, in Table 8.1.

The islands or island groups (except for Krakatoa which is unique because of its youth) are arranged from left to right in the table roughly in order of decreasing degrees of remoteness; the ordering is necessarily somewhat subjective, since many islands have probably been stocked from several "mainlands," over different distances, and from species pools of different sizes and different levels of ecological suitability. Moreover, an island's contemporary "remoteness" has no doubt been increasing continuously as a result of ocean-floor spreading, while colonization has been in progress.

With these reservations in mind, consider Table 8.1. Carlquist (1974) regards the contrast between the Hawaiian Islands and Samoa (columns 1 and 2) as showing that few airborne disseminules can reach very remote islands (such as the Hawaiian Islands) which, as a consequence, have a larger percentage than Samoa of bird-dispersed species.

The absence of airborne species from Cocos-Keeling is explained (Ridley, 1930) by the fact that the islands are made up entirely of coral reef and hence seaborne species dominate its extremely small flora.

Table 8.1 Percentages of indigenous angiosperms with different modes of dispersal in six islands or island groups. (In parentheses below the names of the islands are the total numbers of species; these are the numbers of original immigrants presumed to have been required as ancestors of the present-day native flora.)

	Hawaiian Islands[a] (256)	Samoa[a] (311)	Cocos-Keeling[b] (22)	Galapagos Islands[a] (308)	Galapagos Islands[c] (292)	Christmas Island[b] (104)	Krakatoa[b,d] (140)
Airborne	2	15	0	4	11	9	24
Carried by birds	75	65	23	73	77	49	33
Seaborne	23	20	77	23	12	42	43

[a]From Carlquist (1974).
[b]From Ridley (1930).
[c]From Porter (1976) with pteridophytes excluded.
[d]The species list analyzed for Krakatoa dates from 1919, when the flora was 36 years old.

The contrast between the two columns relating to the Galapagos Islands demonstrates the different conclusions that can be reached by different workers.

In any case, few people who classify island floras on the basis of the modes of dispersal of their species, explain how they infer these modes. For example, a seed or fruit with a parachutelike pappus (many Compositae) or plume (milkweed [*Asclepias* spp], avens [*Geum* spp], cottongrass [*Eriophorum* spp], *Clematis* species, and many more) are not restricted to airborne travel. As Ridley (1930) has pointed out (in what is still, after half a century, perhaps the most exhaustive and authoritative study of plant dispersal), pappuses and plumes can easily become attached to birds' feathers and ensure dispersal by that means. Likewise, the winged seeds and fruits of certain plants (maple [*Acer* spp], ash [*Fraxinus* spp], basswood [*Tilia* spp], and many more), although obviously adapted for airborne dispersal, can "fly" only a very short distance if they are heavy. Therefore they can cover long distances only if, after reaching the surface, they become waterborne on fresh or salt water. The winged fruits of some she-oaks (*Casuarina* spp) can withstand salt water and have almost certainly been spread to oceanic islands in this way. Thus many species have more than one mode of jump dispersal; one mode may carry such a species across wide tracts of ocean to an island, and another mode ensure its spread within the island. Of course the lighter the weight of a winged seed, the farther it can be carried by air currents while still aloft; there is, indeed, no clear distinction between ultra-light-weight seeds with minute winglike dilations, that can float in the air for long periods, and true "dust seeds" (such as those of orchids) which, with the spores of cryptogams, form an important component of the aerial plankton.

Different Forms of Jump Dispersal
and Their Probabilities

The success of different modes of jump dispersal is strongly affected by extrinsic circumstances. Consider again the dissemination of seeds and fruits by the wind. It can take place across open, unforested land, across the sea, and across ice (either ice-covered land or sea ice). Wind dissemination is even more effective over ice than over the sea (Savile, 1968). There are two reasons for this: first, if a seed or fruit is temporarily "grounded" by loss of lift, it can easily resume the journey later if the pause was on ice, but is doomed if it fell in salt water; second, ice, unlike water, does not become choppy and thus tend to slow, through friction, the speed of a rising wind. Therefore wind dissemination is an important mode of plant spread in high latitudes. Savile considers that after the melting of the Wisconsin ice sheets, much of the newly icefree land was colonized by airborne disseminules arriving from the west-north-west (the direction of the prevailing wind) over land covered with packed snow in winter, from refugia in Beringia (see Chapter 4, page 126). Some plants (for example, *Saxifraga tricuspidata*) appear to be adapted to this mode of dispersal since they do not shed their seeds until the time, late in the year, when a hard, dense snow pack has formed.

Wind is equally important as a dissemination agent in high southern latitudes, over the Southern Ocean. It must have contributed to the revegetation, after the Pleistocene glaciations, of the remote sub-Antarctic islands. For example, Godley (1967) has noted the distributions among these islands of two species of rush, *Juncus pusillus* and *J. scheuchzerioides* (see Figure 8.6). Even if the islands existed, and formed

Figure 8.6 Antarctica and surrounding islands to show the distributions of two species of rushes that are presumably post-Glacial immigrants. + *Juncus pusillus*; ◆ *J. scheuchzerioides*. The islands are: 1, Falklands; 2, South Georgia; 3, Crozet; 4, Kerguelen; 5, Tasmania; 6, Macquarie; 7, Auckland; 8, Campbell. (Data from Godley, 1967.)

parts of mainland Gondwana before it broke up, this could not account for the present occurrences on them of conspecific populations. The breakup of Gondwana took place far too long ago for angiosperm species then extant to have persisted until now; there is also the likelihood that most or all of the plants in these islands were destroyed by at least one of the glaciations. However, the northward drift of Australia away from Antarctica had made the Southern Ocean an uninterrupted corridor for strong circum-Antarctic winds, and these were probably stronger at the end of the last glaciation than at present (in the roaring forties, the furious fifties, and the shrieking sixties), since the polar anticyclone may now be less strongly developed. Thus dispersal jumps that might be regarded as improbable at the present time may have occurred frequently in the different circumstances of the past.

Jump dispersal of plants with birds as the agent, when it is examined in detail, has as many interesting complexities as dissemination by the wind. There are three ways in which a bird can transport seeds or fruits: the disseminules, if they are sticky or barbed, can become attached to birds' feathers; or they can be eaten by birds and be carried long distances before being excreted; or they can become imbedded in mud that sticks to birds' feet.

There has been much debate over the probability that these mechanisms, though easy to visualize, actually work. Thus, for internal transport to take place, a disseminule must not be damaged in a bird's crop or by its digestive processes; and if a disseminule is to be carried an appreciable distance, it must not be excreted too soon. Löve (1963) has stressed the improbability of successful plant dissemination by birds. She points out that among migratory birds—the only birds that might carry seeds over a long-distance jump—shore birds are unsuitable because they are nonvegetarian; passerines metabolize too fast, and seed-eating birds digest and hence destroy the seeds they eat. This leaves migrating ducks and geese, in whose crops only hard-shelled seeds would be likely to survive; however, birds about to migrate often refrain from eating for a period in order to fly with low load. External transport of disseminules on a bird's feathers or feet is also less likely than is commonly supposed, according to Löve, because birds usually preen and clean themselves carefully before a long flight; dehydration during flight could also reduce the viability of externally borne disseminules.

Shore birds and waterfowl, whether or not they frequently disseminate flowering plants, are probably responsible for the jump dispersal of many freshwater algae and invertebrates. Proctor, Malone, and DeVlaming (1967) tested the possibility experimentally. They used killdeer *(Charadrus vociferus)* and mallards *(Anas platyrhynchos)* as experimental birds, and the spores and vegetative cells of various algae (including *Chara* species and desmids), and the eggs and adults of several microcrustaceans, as experimental disseminules. The results demonstrate that generalizations about dispersal probabilities are likely to be unreliable; there was found to be considerable variation, between the two bird species and among the various aquatic disseminules, in passage times through a bird from ingestion to excretion. Also the proportions of the disseminules destroyed by digestion increased as passage time increased.

The foregoing paragraphs have discussed the manner in which birds function as the agents for disseminating other organisms. It is worth remarking that birds themselves, even though they can fly, may be very poor dispersers if they are nonmigratory. This is especially true of tropical forest birds, which tend to live sedentary lives in stable habitats. There is a strong tendency for forest birds of tropical islands to be endemic either to a single island or to a closely spaced group of islands. Diamond and May (1976) have noted the absence from islands in the New World tropics of any representatives of several neotropical bird families, in spite of apparently suitable habitats in the islands; similarly, Wallace's line has proved impassable to all the members of many large families of Asian birds.

This brings us to the problem of dispersal probabilities. It should first be observed that for a particular jump dispersal to succeed, not merely must a group of individuals (perhaps in dormant state) traverse the distance to be jumped; these individuals must also manage to establish themselves at the destination, possibly in the face of competition, in sufficient numbers to found a population that can persist.

In the past, biogeographers have engaged in innumerable debates on the probability of this or that species successfully crossing this or that biogeographic barrier. For many such apparent jumps, it would be argued that there were so many all-but-insurmountable difficulties that jump dispersal could not account for a species' presence on an island: the past existence of a land bridge permitting slow diffusion was (it was said) an inescapable conclusion. With the acceptance of continental drift as a means of transporting organisms over long distances, many of the old puzzles were solved. But the problems surrounding the colonization of oceanic islands still remain. That they have been colonized is apparent. Doubt remains over the means by which individual species managed the journey.

Simpson (1952) pointed out the obvious fact that even when a dispersal event was exceedingly unlikely to succeed at a single "trial"—for instance, the carrying of a seed by one bird, on one long-distance migration—the event could become highly probable if enough trials were performed; and the migrations of large flocks of birds every year for millions of years might in some cases be "enough."

This unquestionably correct statement has led to curious misunderstandings. Both Omodeo (1963) and Carlquist (1974) apparently mistook Simpson to mean that all dispersal events, however unlikely, if they are conceivably possible must eventually become likely. Omodeo rejected the idea: he wrote, "If Simpson (1952) is right in believing that highly improbable events become almost certain given a very long period of time, it is also true that if the probability of an event is nil, eternity will not be enough to make it come true." Carlquist accepted it: he states, as a principle of dispersal, that* "Among organisms for which long-distance dispersal is possible, eventual introduction to an island is more probable than nonintroduction." Paraphrased, this amounts to asserting that the probability of every theoretically possible introduction by jump dispersal exceeds 0.5.

It is clear that Simpson's point was unappreciated by these two authors. The point is merely that subjective estimates of dispersal probabilities may be wildly wrong. Omodeo has wrongly imputed to Simpson an absurd belief and has demolished it. Carlquist has imputed to Simpson the same belief and accepted it.

The discussion of long distance jump dispersal in this section should be compared with the discussion of the theory of island biogeography in Chapter 6. Here we have been concerned with a representative few of the myriad interesting though unrelated details that confront any investigator of jump dispersal. Chapter 6 discussed an attempt to generalize the statistical results of dispersals, assuming small islands to be the destinations and time spans to be short, in evolutionary terms. Contemplating the two dissimilar discussions together suggests that theoretical investigations like those of Chapter 6, if applied to dispersal events in all contexts (not only islands) and allowing time spans great enough for appreciable secular changes (geomorphological, climatological, and evolutionary), could lead to breakthroughs in the future. However, when events of such magnitude come to be modeled, there will be endless scope for disagreement over whether the models reveal the signal by excluding the random noise or whether they reveal only noise by excluding the signal [as Sauer (1969) believes the MacArthur and Wilson model has done; see page 176].

In any case, biotic dispersal processes provide a promising field for exploring what level of abstraction can reasonably be striven for in a topic so densely strewn with special cases.

---

*Italicized in the original.

Chapter Nine

DISJUNCTIONS

The disjunct (or equivalently, discontinuous) geographic ranges of many species and higher taxa offer problems that have always intrigued biogeographers. Unraveling the causes of disjunctions has been the motive of a great deal of biogeographic research. Disjunct ranges have been mentioned indirectly in all the first eight chapters of this book but the present chapter deals with the topic explicitly. There are two sections: the first is qualitative and describes, with examples, the various known, or inferred, causes of disjunctions; the second is mathematical and deals with various methods for judging whether the observed range of a taxon is in fact discontinuous when the evidence is ambiguous and intuition unhelpful.

There is no clear, universally accepted definition of the term "disjunction" but this is of no great moment. Obviously, no species-population is continuous in the sense that all its members are in contact; one would not, on that account, say that all ranges were discontinuous. At the other extreme, any terrestrial species whose range is cosmopolitan (or almost cosmopolitan) might be said to have a discontinuous range merely because the world's land surfaces are discontinuous, but the fact would not be of particular biogeographic interest; it would imply only that the organisms concerned (such as rats, mice, and houseflies) spread easily and can thrive almost anywhere.

It is best, therefore, to define *disjunction* subjectively, to mean a range discontinuity that is biogeographically interesting. This is so when a range is divided into two or more well-spaced subranges and raises one or both of these questions: Why is the taxon absent from the area between (or among) its subranges? And how did the taxon come to have spatially separated subranges?

Section 1 deals with clear-cut conspicuous disjunctions about whose reality there can be little doubt. In Section 2 we see that it is sometimes hard to decide, for rare, little-known, or inconspicuous organisms, whether apparent disjunctions are spurious and are merely the result of "disjunctions" in explorers' collecting areas.

1. THE CAUSES OF RANGE DISJUNCTIONS

There are several ways in which the many hundreds of known disjunctions could be classified. They might be classified by region, as Thorne (1972) has done for angiosperm disjunctions. They might be classified by age, that is, according to the time when the range of each currently disjunct taxon changed from a single undivided range to two or more subranges; however, knowledge is insufficient at present to permit such a classification. They might be classified according to their taxonomic ranks, that is, according to whether the disjunct taxon is a species, a section or subgenus, a genus, a family, or an even higher taxon. Such a classification would roughly parallel a classification by age, since the longer a disjunction has existed, the higher will be the rank of the taxa in the separate subranges; this correspondence would apply only within defined larger groups (phyla or classes, say) because of their different evolutionary rates.

Lastly, disjunctions can be classified by cause, as is attempted below. In schematic form, the classification is as follows:

1. Separation into parts of a once-continuous range because of:
   a. Geomorphological changes.
   b. Climatic changes.
   c. Evolutionary differentiation plus migration.
2. Establishment of new subranges by long-distance jump dispersal owing to:
   a. Natural jump dispersals.
   b. Jumps aided by human agency.

We now consider these categories in turn. Some examples mentioned in earlier chapters are recalled and some additional examples are given. (Probably all would be disputed by at least a few biogeographers.)

1a. Geomorphological Disjunctions.   Many of the geologically oldest and taxonomically highest-ranking disjunctions have been caused by the fragmentation of the continents and the drifting away from each other of

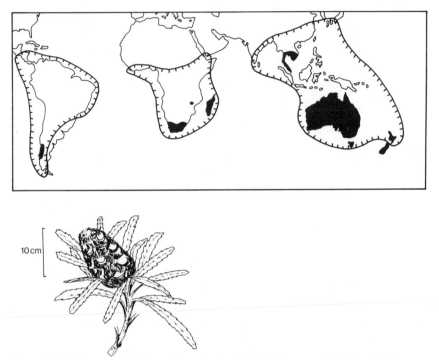

Figure 9.1 The ranges of the angiosperm families Restionaceae (black) and Proteaceae (outlined). (Redrawn from Hutchinson, 1926, and Good, 1974.) Lower left: *Banksia serrata,* an Australian member of the Proteaceae.

the parts. Some examples given in Chapter 2 may be recalled. For instance, the disjunction, between Australia and South America, of the subclass Marsupiala is believed to be due to the breakup of Gondwana (see page 57). So is the disjunction in the ratites (superorder Ratitae), the flightless running birds, which are represented by different orders or families in the various Gondwana fragments (page 53). Among angiosperms, the disjunct ranges of the Gondwana families Proteaceae (page 47) and Restionaceae are famous (see Figure 9.1).

Continental drift has joined continents as well as separating them, and the joining of continents has broken the continuity of the sea. For example, taxa of the marine fauna common to the tropical east Pacific and the Caribbean had continuous ranges before the Isthmus of Panama formed a barrier; now they are disjunct; the "geminate" species-pairs having one member on each side of the barrier (page 151) are disjunct subgenera.

The submergence of land bridges, independently of the drift of tectonic plates, can disrupt terrestrial ranges. Thus the possible submergence of a land bridge spanning part of the North Atlantic may have contributed to the disjunctions in some plant taxa (species and subgenera, or sections) with amphiAtlantic ranges (page 222). Conversely, the emergence of the Bering land bridge at various times has rendered disjunct the ranges of marine taxa that are now found in Bering Strait and the seas to north and south of it (page 133).

The submergence of coastal plains, because of the eustatic rise in sea level resulting from the melting of the Pleistocene ice sheets, may often have created gaps in once-continuous ranges. Probably there are many disjunctions from this cause. An Australian example, in which two *Eucalyptus* species have become disjunct, was described earlier (page 112 and Figure 4.2).

1b. Climatic Disjunctions. Geomorphological change has usually been accompanied by climatic change. Continental drift has shifted land areas into warmer or cooler latitude zones. Crustal warping has converted hot lowlands into cool highlands and vice versa. Climatic deterioration can make the center of the range of a widespread taxon uninhabitable for it and leave the extremities as disjunct subranges.

The best-known and most numerous examples are provided by those plant taxa with subranges in eastern Asia and eastern North America. The communities they belong to, together with patches of forest in western North America and western Eurasia, are remnants of the great mesophytic Arcto-Tertiary forest that covered most northern hemisphere land including Beringia (see page 132). The cooling that destroyed the Beringian section of the forest was caused in part by the worldwide cooling trend that has been in progress since the early Tertiary and in part by the northward drift of Beringia into higher latitudes.

A large number of disjunct Arcto-Tertiary forest taxa are known. Li (1952) has discussed those now occurring in eastern Asia and eastern North America. They have also been considered by Wood (1972) who points out that there is no reason to suppose that the many disjunctions are all of the same age; no doubt breaks in once-continuous ranges occurred in different taxa at different times in the more than 60 m.y. of the Tertiary and Pleistocene. Also, the surviving communities have probably shifted north and south, perhaps over 1000 km and more, as the Pleistocene ice sheets expanded and contracted.

There is considerable debate over the taxonomic status of the disjuncts. Some authors consider that many of the taxa common to eastern Asia and eastern North America are conspecific. According to Li (1952), however, the floristic relation between the two areas "is primarily generic, not specific," though among Pteridophytes, he concedes that the populations of *Onoclea sensibilis* (sensitive fern) and *Osmunda claytoniana* (inter-rupted fern) in the two regions may be conspecific. Disjunct angiosperm genera are numerous. Besides those listed in Chapter 4, it is worth mentioning seven more, chosen because of their familiarity to gardeners as ornamentals: *Liriodendron* (tulip tree), *Magnolia*, *Hamamelis* (witch hazel), *Catalpa*, *Wisteria*, *Pachysandra*, and *Astilbe*.

It should be recalled (page 132) that insects (and presumably other invertebrates) have also persisted in the relict tracts of Arcto-Tertiary forest. Thus Linsley (1963) has given several examples of disjunct ranges in the Cerambycidae (Longhorn beetles) and as an illustration the range of the genus *Calloides* is shown in Figure 9.2. It is found in western North American relicts of the forest as well as in the two larger tracts discussed above.

Climatic change has been the cause of disjunctions in many other parts of the world. For example, it is believed (Axelrod, 1970, 1972) that many tropical rain forest trees had ranges that encompassed the whole of tropical Africa in the Cretaceous and early Tertiary. Then (see page 43) upwarping of the crust by as much as 3000 m caused the climate in the center of the continent to become cooler and drier, and many species became globally or locally extinct; local extinction, with survivors per-

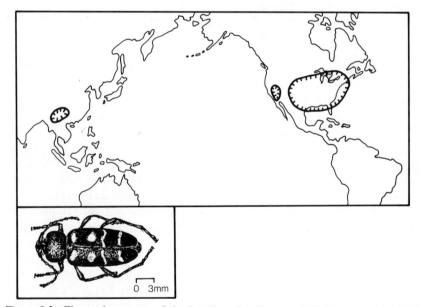

Figure 9.2 The modern range of the Longhorn beetle genus *Calloides*, a genus of the Arcto-Tertiary forest. (Redrawn from Linsley, 1963.) Lower left: *Calloides nobilis*.

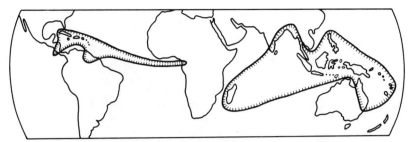

Figure 9.3 Range of the lowland tropical rain forest genus *Hernandia* which probably (Axelrod, 1972) disappeared from most of Africa when land levels rose and the interior of the continent became cooler and drier. (Redrawn from Hutchinson, 1926.)

sisting only in Madagascar and coastal West Africa, may account for the trans-African disjunction of the genus *Hernandia* (see Figure 9.3)

Increasing aridity in inland North America, which led to the development of prairie grassland as climax vegetation, is probably responsible for the present disjunct range, within the continent, of the mammalian family Talpidae (the moles) (see Figure 9.4). According to Burt (1958) moles occurred all across the continent in the early Tertiary. Now they are confined to two widely separated regions; there are two genera (*Neürotrichus* and *Scapanus*) in the West, and three (*Condylura, Scalopus,* and *Parascalops*) in the East.

Among plants, there are large numbers of amphitropical American disjuncts. These are taxa having two widely separated subranges (which may themselves be further subdivided), one in North America and the other in South America. Most of them probably owe their present pattern to long-distance jump dispersal and are therefore discussed below. However some species (or groups of closely related species) of the woodland genera *Osmorhiza* and *Sanicula* (both of the family Umbelliferae) have this pattern (Constance, 1963) and it is likely (Raven, 1963) that they spread by slow diffusion along connected mountain chains, and subsequently became extinct in the centers of their ranges, which were long and narrow. *Osmorhiza chilensis*, whose range is shown in Figure 9.5, is an example. The presence of tropical montane species of the genus more or less bridging the disjunction suggests that the pattern is not the outcome of a long-distance jump (Raven, 1963).

Some marine disjunctions stem from changes in the climate of the sea. An interesting example, due to Bousfield and Thomas (1975), was described in Chapter 4 (page 114). A rise in sea level over the Atlantic coastal shelf of North America during the Holocene appears to have affected the thermal stratification of the water, and hence the environment

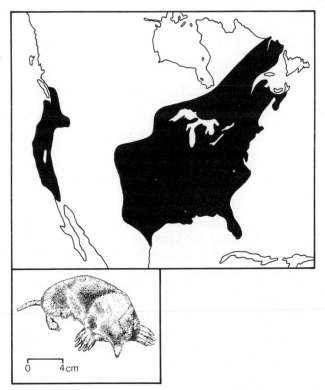

Figure 9.4 The disjunct range in North America of the family Talpidae (the moles). (Redrawn from Burt, 1958.) Lower left: *Scapanus orarius,* one of the western moles.

of benthic organisms. Some species have become excluded from parts of their former ranges, leaving the survivors disjunct.

1c. Evolutionary Disjunctions. When a pair of sister subspecies or species has become differentiated on opposite sides of the area occupied by their common ancestor, and the ancestor then becomes extinct, the descendent taxa form a disjunct pair. Sometimes they resemble each other so closely that they are classed as one species. Raven (1963) believes that this mechanism accounts for some of the amphitropical American disjuncts of desert habitats, plants that are found now in North America in the Sonoran and Chihuahuan deserts and in South America in the Chilean and Peruvian deserts. Among examples he lists are the woody genera *Ficus* and *Acacia.*

Opinion varies on how recent the differentiation must be for two descendent populations from a common ancestral population to be treated

Figure 9.5 The range of the amphitropical disjunct *Osmorhiza chilensis* (family Umbel-liferae), a herbaceous, woodland plant. (Redrawn from Constance, 1963.)

as true disjuncts. Thus Solbrig (1972) treats as only apparent the disjunction between two species of *Prosopis*, *P. juliflora* (mesquite) of North American deserts and the very similar *P. chilensis* of South American deserts. He considers that their great morphological similarity has resulted from convergent evolution; both species, in their descent from a rather distant common ancestor, have evolved matching adaptations to matching arid environments; but their relationship to each other is less close than that of either to morphologically dissimilar *Prosopis* "sisters" that have remained in mesic environments and not been selected for xeromorphy.

It is likely that at least some of the known cases of disjunctions in marine plankton taxa can be classed as evolutionary disjunctions. The

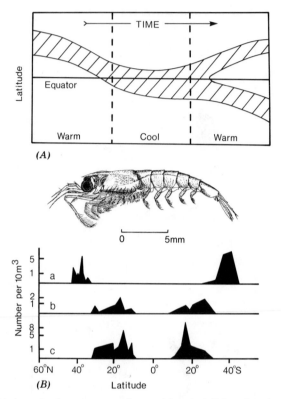

Figure 9.6 (*A*) A mechanism to account for amphitropical disjunctions in marine plankton species. (*B*) The densities, at different latitudes in the mid-Pacific, of three species of euphausid crustaceans: *a*, *Thysanoessa gregaria*; *b*, *Stylocheiron subhumii*; *c*, *Euphausia brevis*. (Redrawn from Johnson and Brinton, 1963.) Inset: *Thysanoessa gregaria*. (Redrawn from Thomson and Murray, 1885.)

probable course of events is shown diagrammatically in Figure 9.6*A*. At its simplest, the model assumes that at some past time an ancestral taxon had a nondisjunct range in one or other hemisphere; the climate then cooled, and in order to remain in water of tolerable temperature, the organisms were forced to migrate into lower latitudes until their range bracketed the equator. Later, when the climate warmed again forcing a poleward migration, the range split in two with some descendants moving north and others moving south. Johnson and Brinton (1963) have given some examples of amphitropical range disjunctions in the Pacific that may have been caused by this mechanism; the organisms are euphausid crustaceans. Figure 9.6*B* shows the latitudinal belts occupied by three of these species along (approximately) a north-south meridian in the mid-Pacific. If their range disjunctions are the result of the mechanism outlined above, one would expect the northern and southern populations of each species to diverge from each other evolutionarily; in the species shown here, they have not done so yet. In *Thysanoessa gregaria*, discussed in detail by Johnson and Brinton, the northern and southern populations are morphologically indistinguishable and genetic differences, if there are any, are not yet discernible. In both hemispheres, the subranges of this species lie between the 7°C and 11°C isotherms (at 200 m depth); but salinities are more variable in the northern than in the southern subrange, suggesting that salinity is less important than temperature in determining the species' range. If the model does explain the range disjunctions of these species, it follows that their ranges are determined by conditions in the epipelagic (photic) zone, at depths of less than about 200 m; in this zone, conditions are closely coupled to those in the overlying atmosphere and tend to vary smoothly with latitude. In contrast, at greater depths—in the mesopelagic zone and at still greater depths—lie the "water masses" whose internal temperatures and salinities are fairly constant over extensive areas (see Chapter 5, page 140). If the geographic ranges of plankton species are coincident with water masses, as many oceanographers believe, then the model diagrammed in Figure 9.6 is presumably not correct. McGowan (1971) has argued that worldwide cooling would leave patterns of oceanic circulation unchanged and would not lead to the shifting or coalescing of water masses implied by the middle section of the diagram.

2a. Natural "Jump" Disjunctions. Long-distance jump dispersal, or a combination of long-distance jumps and diffusion, has no doubt been responsible for a great many disjunctions, especially in plants. For example, Raven (1963) has listed approximately 160 plant species (or species groups) of temperate and cool temperate environments in the Americas that have amphitropically disjunct ranges, and he gives very strong evidence for supposing that these disjunctions have resulted from jump dispersals in all but a few exceptional cases (two of these exceptions are woodland members of the family Umbelliferae, mentioned under **1b** above).

The reasons for believing these disjunctions to have been caused by relatively recent jump dispersals are as follows. The disjuncts are all herbaceous plants of open habitats, such as grasslands or seashores, and would therefore be more likely to establish themselves successfully after a long-distance jump than would plants of closed communities such as woodlands. All have disseminules that are well adapted to jump dispersal. Apart from the disjuncts, the floras of the two regions are very dissimilar and have probably been distinct since the Cretaceous or earlier. The disjuncts are therefore disharmonic; thus those of northern origin (which form the great majority) are "out of place" in their southern, "satellite" subranges and occur there without all the many other plants and insects customarily associated with them in their "home" subranges in the north. Lastly, all are self-fertile. Indeed a disjunct species is often the only self-fertile species in a genus whose other members are all self-sterile and do not have disjunct ranges. Obviously a species that can fertilize itself is much better equipped to become established after a long-distance jump than is a self-sterile species since a single individual is all that is needed to found an immigrant population in the "satellite" subrange.

The evidence suggests that many of these amphitropical disjunctions are as recent as the Pleistocene. The conditions for successful jump dispersal were probably especially good when glaciations were abating and ice sheets were melting. There would then be extensive areas of suitable habitat for these plants of open environments; this would ensure large populations and an abundant supply of seed in the source region and it would also ensure large tracts of easily invaded land in the satellite region. Furthermore, immediately after a glaciation climatic zones were still telescoped; any climatically matched pair of zones with one zone in the northern hemisphere and the other in the southern would have been closer to each other geographically at the peak of a glaciation than at any other time.

Three examples of these amphitropical disjuncts, chosen for mention because of their large ranges in North America are: *Hordeum jubatum* (foxtail barley), *Myosotis verna* (an annual forget-me-not), and *Linaria canadensis* (an annual toadflax). Notice that two of these are annual plants. Their seeds can compete on equal terms with those of indigenous annuals once their long-distance jumps have been achieved.

Another set of disjunct plants whose biogeography has been closely studied are the amphi-Atlantic taxa. Data concerning them, and detailed range maps, were presented by Hultén (1958). They have been discussed here in Chapter 7 (page 222). It will be recalled that the disjunctions entailed jump dispersal at least across Davis Strait between Baffin Island and Greenland (Lindroth's opinion) or across Denmark Strait between Greenland and Iceland (Löve's opinion).

Jump dispersal must also account for the enormously wide disjunctions, among remote islands in the Southern Ocean, of the rushes *Juncus pusillus* and *J. scheuchzerioides* as described by Godley (1967) (see page 262 and Figure 8.6).

Another class of disjunct plants deserves special mention. These are species belonging to genera in which some species are homostylous and others heterostylous. In a homostylous (or monomorphic) plant the arrangement of styles and stamens is the same in every flower. But the flowers of a heterostylous (or dimorphic) species are of two* kinds: in the flowers of some individuals, the styles are longer than the stamens, and in other individuals the arrangement is reversed. In genera with some homostylous and some heterostylous species, the homostylous species are usually self-fertile and the heterostylous species, whose two forms of flower were evidently selected to facilitate cross-pollination, are self-sterile. When species in such genera are disjunct, it is found almost without exception that only the homostylous species are disjunct; the heterostylous species, in contrast, have unbroken ranges. The disjunctions in homostylous species have presumably resulted from the establishment of remote satellite subranges by jump dispersal; in each case, establishment succeeded because a single self-fertile plant was able, by itself, to found a satellite population. The topic has been discussed by Baker (1959); two good examples he mentions are the genera *Primula* (primrose) and *Armeria* (thrift).

The examples of jump-caused disjunctions described above have all been of plant species. Among terrestrial species of all kinds (as opposed to taxa of higher rank), such disjunctions are much commoner in plants than in animals. Raven (1963), in his discussion of amphitropical American plants, commented that there were no known insect species with comparable patterns; a few amphitropical genera and families of insects are known, but their disjunctions may have arisen because of evolutionary divergence from a common ancestor. In only one insect genus is it quite probable that the disjunction was caused by jump dispersal. This is the praying mantis genus *Brunneria*. Several species of the genus occur in South American grasslands, in Brazil, Paraguay, and Argentina, and the species-populations contain female and male individuals in equal proportions. But in the satellite subrange of the genus in North America, in prairie grassland from Texas to North Carolina, the genus is represented by a single species, *B. borealis*, which consists entirely of females (Rehn, 1948). The all-female species-population may be a clone descended from a single ancestor that chanced to survive an accidental jump dispersal. This is not the only plausible explanation, however. Rehn considered the disjunction to have come into existence when the extensive dry grasslands of the Pliocene, which at one time provided a continuous habitat for

---

*There are also a few species with trimorphic flowers, having long, medium, and short styles respectively. See Briggs and Walters (1969).

the genus, were fragmented by the spread of mesic vegetation in the more humid Pleistocene. If this is the explanation of the disjunct range of the genus *Brunneria*, it should be classed as a **1b** disjunction (see page 267).

Now consider the marine realm. Jump dispersals by shelf organisms across the deep ocean can lead to disjunctions in exactly the same way as jump dispersals of terrestrial organisms across the sea or tracts of ecologically unsuitable land. However, such disjunctions are probably less common than in land plants; the motile, migrating stages of most sedentary animals of the shelf benthos—mollusks, echinoderms, barnacles, and the like—are immature larval stages and are much more delicate and short-lived than the dormant disseminules (spores and seeds) of vascular plants. Examples of mollusk species with disjunctions that were shown (Scheltema, 1971) to have resulted from the transoceanic dispersal of veligers are the tun shell *Tonna galea* and the dogwinkle *Thais haemastoma* which occur on both the eastern and western shores of the Atlantic (see page 170 and Figure 5.10).

2b. Man-made Disjunctions. Whether disjunctions resulting from human action should be set apart from "natural" disjunctions and treated as "unnatural" is a matter of choice. Man-made disjunctions are of two kinds, deliberate and unwitting. Neither is discussed here, but it is worth remarking that in studies of "natural" biogeography (uninfluenced by human actions) it is easy to be misled and to mistakenly regard as natural what is, in fact, a man-made disjunction. The plants of coastal regions are especially likely to be misleading; one of the most effective means of transoceanic jump dispersal of seeds is in ships' ballasts. For example, no fewer than 257 of the plants of Newfoundland—23% of the total flora—are European species. Their seeds were probably carried with the stony ballast picked up on European shores to ballast ships returning to Newfoundland after delivering cargoes of fish caught on the Grand Banks of Newfoundland (Lindroth 1957).

This section should not end without mention of "spurious disjunctions," that is, range discontinuities that are apparent rather than real. Some amphitropical "disjunctions" in marine plankton organisms found in surface waters at high latitudes may be of this kind. Their ranges are continuous but, because they cannot endure warm water, they are found at progressively greater depths the lower the latitude (see Figure 9.7). Thus an apparent disjunction appears in their ranges if surface water plankton only is mapped. Two examples already mentioned (Chapter 5, page 168) are the scyphozoan genera *Atolla* and *Periphylla*.

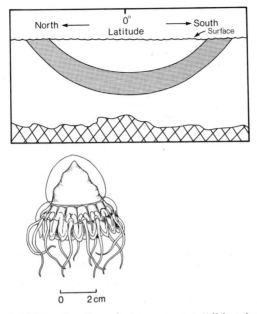

Figure 9.7 The possible explanation of some apparent "disjunctions" in cold-water species of the marine plankton. In low latitudes they occur only at depths rarely sampled. Lower left: *Periphylla hyacinthina*. (Redrawn from Mayer, 1910.)

*2. TESTING FOR "CRYPTIC" DISJUNCTIONS

The preceding section discusses clear, unmistakable disjunctions, in which there is no doubt as to the reality of the gap (or gaps) separating the two (or more) subranges of a taxon. This section is concerned with what might be called "cryptic" subranges and their recognition. Whereas earlier discussions dealt with subranges that were as large as half a continent in extent and were separated by gaps the size of oceans, we now consider patterns on a much smaller scale: subranges and gaps whose linear dimensions are of the order of tens or hundreds of kilometers.

To introduce the subject, consider Figure 9.8 which maps the occurrences of an imaginary littoral species along an imaginary coast; for simplicity, we begin with species whose ranges can be represented, approximately, by lines (not necessarily straight); the maps are called "linear maps." The argument is subsequently generalized to ranges spread over two-dimensional areas shown on "area maps." Map A in the figure shows, with solid symbols, points on the coast where the species is known to occur. On seeing such a map, a cautious observer would

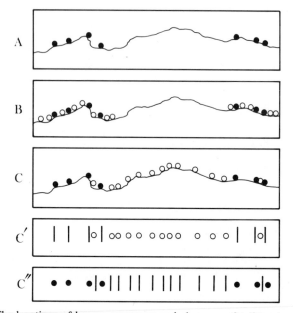

Figure 9.8 The locations of known presences and absences of a littoral species along a shore. Maps A, B, and C show the shoreline itself with presences (●) and absences (○). Maps C′ and C″ are stylized versions of map C. Map C′ shows the absences in blocks defined by substituting partitions for the presences. Map C″ shows the presences in blocks defined by substituting partitions for the absences.

concede that there may be a disjunction; an incautious observer would instantly conclude that there is certainly a disjunction. Now notice that map A shows only known occurrences ("presences"). It does not show, as do maps B and C, points on the coast where the species was sought but not found ("absences"). Absences are of two kinds: "chance" absences, or locations where the species might have occurred but happened to be missed; and "true" absences, or locations from which the species is absent for biological or historical reasons. There is no way of recognizing these two kinds of absences. One must rely on probability theory to yield probable inferences.

Obviously, conclusions about the range of the species shown in map A are strongly affected by the locations of absences. Consider maps B and C. The presences (solid symbols) shown on them are in the same locations as in map A; but the patterns of absences (hollow symbols) are entirely different in the two maps. If map B is the "total" map, there is clearly no reason to suspect a disjunction. The "disjunction" is in the pattern of sites examined; presences and absences appear to be mingled at random, and the absences are merely locations within the species' range where it chanced not to be found. The "artificial disjunction," due to the disjunction in examined sites, may be merely an accident, the result of an arbitrary choice of collecting sites by the field observer. Or it may reflect environmental conditions; thus if the species were a barnacle, say, an observer would not attempt to find it on sandy stretches of shore; and rocky areas, where the species would be expected, might themselves be disjunct.

If map C is the total map, clearly there *is* reason to hypothesize a true disjunction, assuming that the observer searched for the species only in "reasonable" places. To test the hypothesis, it is necessary to determine the probability of obtaining such a pattern (or one even more suggestive of disjunction) under the null hypothesis that there is no disjunction.

Before considering how this probability may be calculated, two points should be noted:

1. The assertion that map B appears to contain no disjunction has different implications depending on whether the unexamined middle stretch of shore was left unexamined for reasons unconnected with biogeography and ecology, or was excluded because it contained no suitable habitat for the species concerned. In the former case one might reasonably suppose that the species would be found there if a search were made: the gap is truly "artificial." In the latter case it is debatable whether a stretch of shore that is uninhabitable by the species should or should not be treated as a "true" gap. The answer is unavoidably subjective and is influenced by the size of the gap and the dispersal powers of the species.
2. The assertion that map C *does* appear to contain a disjunction is likewise open to different interpretations depending on circumstances. If the long row of absences in the middle stretch of shore coincides with unsuitable habitat, we could accuse the field observer of concealing evidence by not showing that the mapped absences were at sites having an unsuitable environment for the species. But if environmental conditions are more or less uniform along the whole length of mapped coast, then the map does indeed suggest (subject to testing) a true disjunction. Notice, however, that matters are seldom as clear-cut as this in practice. Environmental conditions are rarely uniform (even "more or less uniform"). Instead of two sharply contrasted habitats (such as pure rock versus pure sand), we are likely to find that several environmental factors vary continuously. Our lack of knowledge of the species' ecological tolerances

may sometimes prevent us saying whether absences such as those in map C are sites from which the species is ecologically excluded (in which case the map does not differ in principle from one version of map B) or are "surprising" absences indicative of true disjunction.

The foregoing comments emphasize the careful consideration that must always come between the decision to do a statistical test and the doing of it. The chief difficulty lies in selecting an appropriate null hypothesis. Three (at least) versions are possible. It is convenient to give descriptive names to the tests of these three somewhat different hypotheses. (All three are described as one-tail tests.)

The *runs test* tests whether the presences and absences are randomly mingled with each other against the alternative that they are grouped relative to each other.

The *blocks test* tests whether the absences are randomly scattered among the presences, against the alternative that they form a compact group (or groups) among the presences. (Or, if desired, it does the converse test, defined by exchanging the words "presence" and "absence" in the definition.)

The *spread test* tests whether the absences are spread out along as extensive a strip of land as the presences, against the alternative that they are confined to a shorter strip. (Or the converse test, defined by exchanging the words "presences" and "absences.")

For clarity the tests are described below in the following order: the runs test and then the blocks test, applied to linear maps; the blocks test and then the runs test applied to area maps; lastly, the spread test.

To reiterate, before embarking on details: although the tests are easy to do, the problem of choosing which of them is appropriate in any given situation is not at all straightforward. Whenever a test is to be applied, there will probably be much scope for debate as to which test should be used.

It will also be noticed that the "converse" versions of the blocks test and the spread test, mentioned above in parentheses, test whether presences (rather than absences) form a compact group. One might not, spontaneously, regard them as tests for disjunctions. However they do test for disjunctions in circumstances that are frequently encountered. If the locations of the presences and absences of a species are mapped in a region that lies entirely inside the known boundaries of the species' range, and if it turns out that the presences form a spatially restricted subset of the locations examined, then it is reasonable to conclude that the species-population as a whole is fragmented into local populations or demes (see page 84 and Figure 3.6).

Linear Maps: The Runs Test and the Blocks Test

As applied to linear maps, the runs test is well known and is described in books on probability theory such as Feller (1968).

Suppose, for concreteness, the sequence of presences and absences in map C of Figure 9.8 is to be tested. The sequence is as follows:

● ● ● ○ ● ○ ○ ○ ○ ○ ○ ○ ○ ○ ○ ○ ● ● ○ ●

Let the numbers of presences and absences be M and N respectively. (In the example, $M = 7$ and $N = 13$.) Let the number of runs (uninterrupted sequences of symbols of either kind) be r; (in the example, $r = 7$).

We first require the probability, say $p(r)$, that there would be exactly r runs if M presences and N absences are randomly mingled. We then find

$$P(r) = \sum_{j=2}^{r} p(j)$$

which is the probability that there would be r or fewer runs (obviously, there cannot be fewer than 2). Then if $P(r)$ is small, say 5% or less, we can reject the null hypothesis at the 5% significance level and conclude that the presences and absences occur in unexpectedly long runs and the runs of absences represent true disjunctions (one or more). But if $P(r)$ is large, there is no reason to reject the null hypothesis and we conclude that the absences are due only to chance and not to disjunctions.

The expression for the probability $p(r)$ depends on whether r is even or odd. Suppose, first, that it is even and put $r = 2k$. That is, there are k runs of presences and k runs of absences.

Note first that the total number of distinguishable arrangements of the presences and absences is

$$\binom{M + N}{M} = \binom{M + N}{N} = \frac{(M + N)!}{M! \; N!}.$$

Now let $X(k; A)$ be the number of distinguishable ways in which A indistinguishable objects can be partitioned into k batches. It is*

*To see this, visualize the A objects arranged in a row. There are $A - 1$ spaces between pairs of adjacent objects. They can be separated into k batches, each containing one or more objects, by putting $k - 1$ "partitions" into the spaces with not more than one partition in any one space. The number of ways in which this can be done is

$$\binom{A - 1}{k - 1}.$$

$$X(k; A) = \binom{A - 1}{k - 1}, \qquad k \le A. \tag{9.1}$$

Then the number of ways of arranging M presences (objects) in k runs (batches) and, independently, N absences in k runs is

$$2 X(k; M) X(k; N) = 2 \binom{M - 1}{k - 1}\binom{N - 1}{k - 1}.$$

The factor 2 occurs because, though runs of the two kinds must alternate, there are two ways of choosing which shall be first. Then

$$p(2k) = \frac{2 \binom{M - 1}{k - 1}\binom{N - 1}{k - 1}}{\binom{M + N}{N}}. \tag{9.2}$$

Now suppose r is odd. Put $r = 2k + 1$. That is, there are k runs of presences and $k + 1$ of absences, or vice versa. By arguments similar to those above it is easy to show that

$$p(2k + 1) = \frac{\binom{M - 1}{k - 1}\binom{N - 1}{k} + \binom{M - 1}{k}\binom{N - 1}{k - 1}}{\binom{M + N}{N}}. \tag{9.3}$$

Here the component

$$\binom{M - 1}{k - 1}\binom{N - 1}{k}$$

is the number of ways of arranging the presences and absences to give k runs of presences and $k + 1$ runs of absences; and the component

$$\binom{M - 1}{k}\binom{N - 1}{k - 1}$$

is the number of arrangements giving the converse pattern.

Finally, as already explained,

$$P(r) = \sum_{j=2}^{r} p(j)$$

is the probability we require, that of obtaining as few runs as observed or fewer.

In the example, $r = 7$ and $P(r) = 0.095$. Thus in spite of the apparent "disjunction," a pattern such as that observed is not particularly improbable even if the presences and absences are randomly mingled with each other.

Now consider the blocks test. It is a statistical test often used in other contexts and a mathematically rigorous account of the theory can be found in Wilks (1962: page 234).

Look at map C in Figure 9.8 again and suppose the presences are replaced by partitions (or "cuts" to use the mathematical term) which divide the region into "blocks" (mathematically, they are "statistically equivalent blocks" or "sample blocks"). The partitioning is shown in map C′ which is a stylized version of map C modified in this way. Note that the regions to the left of the leftmost cut and to the right of the rightmost cut, as well as the regions between cuts, are all "blocks." Therefore if there are M cuts there are $M + 1$ blocks.

Now observe how the N absences are allocated to the $M + 1$ blocks. The null hypothesis is that the absences have been randomly and independently assigned to the blocks. Each block, regardless of its size, has an equal chance of receiving each absence. The justification for this at-first-sight surprising assumption is as follows. Suppose a collector were searching an unexplored region for a species with known habitat requirements. He would first concentrate on finding "reasonable" sites to examine, that is, sites with suitable habitat, and would only then note whether an examined site was a presence or an absence. Reasonable sites are unlikely to be evenly spread on the ground (for example, rocks suitable for a particular barnacle species are not likely to form an equidistant row) and, obviously, the density of absences cannot exceed the density of examined sites. The null hypothesis can be rephrased thus: the probability that an examined site is an absence is the same for all examined sites even though the density of examined sites varies from place to place. It looks true of map B in Figure 9.8, for instance.

Now count z, the number of empty blocks, that is, blocks not containing absences; in the example, $z = 5$.

Next calculate (as shown below) $q(z)$, the probability that exactly z blocks would be empty given the null hypothesis. Then obtain

$$Q(z) = \sum_{j=z}^{M} q(j)$$

which is the probability that z or more blocks would be empty; (obviously z

attains its maximum vaue, M, when all absences are in one block). If $Q(z)$ is small, say less than 5%, we may assume that the absences are grouped into unexpectedly few of the blocks.

The probability $q(z)$ can be derived in two ways. Note first that

$$q(z) = \frac{\begin{pmatrix} \text{the number of ways of} \\ \text{choosing } z \text{ of } M + 1 \\ \text{blocks to be empty} \end{pmatrix} \begin{pmatrix} \text{the number of ways of} \\ \text{putting } N \text{ absences in} \\ M + 1 - z \text{ blocks so} \\ \text{that none is empty} \end{pmatrix}}{\begin{pmatrix} \text{total number of arrangements of } N \text{ absences} \\ \text{and } M \text{ presences} \end{pmatrix}}.$$

Using the expression introduced in (9.1), it is seen that the second factor in the numerator is

$$X(M + 1 - z; N) = \binom{N - 1}{M - z}.$$

Therefore

$$q(z) = \frac{\binom{M + 1}{z}\binom{N - 1}{M - z}}{\binom{M + N}{N}}. \tag{9.4}$$

We now derive $q(z)$ by another method, which will be needed later in the discussion of area maps. Observe that the existence of z empty blocks implies that there are $M + 1 - z$ nonempty blocks and these contain "runs" of absences in the terminology of the runs test. Thus, denoting the number of runs of absences by k, $k = M + 1 - z$. It can now be seen that

$$
\begin{aligned}
q(z) = \quad &\Pr\ [k \text{ runs of absences and } k + 1 \text{ runs of presences}] \\
&+ \Pr\ [k \text{ runs each of absences and presences}] \\
&+ \Pr\ [k \text{ runs of absences and } k - 1 \text{ runs of presences}].
\end{aligned}
$$

Using (9.2) and (9.3), this is found to be

$$q(z) = \frac{\binom{N - 1}{k - 1}\left\{\binom{M - 1}{k} + 2\binom{M - 1}{k - 1} + \binom{M - 1}{k - 2}\right\}}{\binom{M + N}{N}}.$$

$$= \frac{\binom{N - 1}{k - 1}\binom{M + 1}{k}}{\binom{M + N}{N}}$$

Putting $k = M + 1 - z$, this is seen to be identical with (9.4).

Returning to the example in map C' we have, as before, $N = 13$ absences and $M = 7$ presences (now partitions). The number of empty blocks is $z = 5$ and

$$Q(z) = \sum_{j=5}^{7} q(j) = 0.052.$$

The probability of obtaining the observed map is thus less under the blocks test hypothesis than under the runs test hypothesis.

It is interesting to do the converse blocks test. The hypothesis being tested is now that presences are randomly and independently assigned to blocks bounded by partitions put in place of the absences (see map C'' in Figure 9.8). Write z' for the number of empty blocks and calculate $Q'(z')$, which is obtained in the same way as $Q(z)$ except that M and N are interchanged. The result is $Q'(z') = 0.336$.

It is worth showing diagrammatically how the different components of the probabilities $p(r)$, $q(z)$, and $q'(z')$ are combined. This is done in Figure 9.9. Assume that the sequence of $M = 4$ presences (●) and $N = 7$ absences (○) shown at the top of the figure is being tested.

Consider runs first. The number of runs of presences, which we now call k_p, could range from 1 to 4 (in fact, $k_p = 2$). The number of runs of absences, k_a, could range from 1 to 5 (in fact, $k_a = 3$). One can therefore display, in a 5×4 array, the probabilities of all possible combinations of k_a and k_p. These probabilities exceed zero only in the cells of the array with solid outlines, that is, those in which $|k_a - k_p| = 1$. The number of runs of both kinds, $r (= k_a + k_p)$ can range from 2 to 9 ($= 2M + 1$); in fact, $r = 5$.

Now consider the four arrays in the figure. The top left array shows, in the (i, j)th cell, the number of ways of obtaining a sequence with $k_a = i$ and $k_p = j$. To obtain probabilities, these values are divided by

$$\binom{M + N}{M} = 330.$$

Thus [compare (9.3) and (9.4)],

$$\Pr[i = k_a, j = k_p] = \frac{\binom{N - 1}{k_a - 1}\binom{M - 1}{k_p - 1}}{\binom{M + N}{M}}.$$

The top right array shows how the probabilities are combined to give values of $p(r)$. The bottom left and bottom right arrays show, respec-

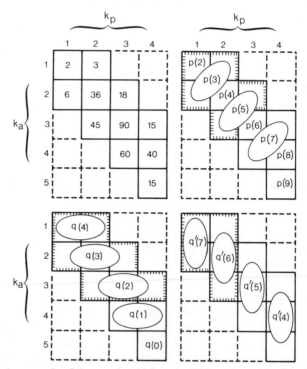

Figure 9.9 An example of the way in which probabilities are combined to give values of $p(r)$, $q(z)$, and $q'(z')$ when there are k_p runs of presences (●) and k_a runs of absences (○). The sequence under test, with $M = 4$ presences and $N = 7$ absences, is shown at the top. See text for further details.

tively, how the probabilities are combined to give $q(z)$ (with $z = M + 1 - k_a$) and $q'(z')$ (with $z' = N + 1 - k_p$). The groups of cells with hatched outlines show the cell entries that must be summed to give, respectively, the required probabilities for testing the sequence at the top of the figure.

For the runs test, $r = 5$; we require $P(r) = \Sigma_{j=2}^{5} p(j) = 0.333$ (top right array).

For the blocks test with partitions through presences, $z = 2$; we require $Q(z) = \Sigma_{j=2}^{4} q(j) = 0.652$ (bottom left array).

For the converse blocks test, with partitions through absences, $z' = 6$; we require $Q'(z') = \Sigma_{j=6}^{7} q'(j) = 0.279$ (bottom right array).

It will now be realized how strongly the final inference depends on the hypothesis that was chosen for testing. This is decided by circumstances and by the investigator's interests and preconceptions. With no advance opinion on whether each examined site will turn out to be a presence or an absence, one would do the runs test, which is a symmetrical test. If one expected examined sites to be presences, and regarded an absence as a cause for interest and surprise, $Q(z)$ would be the appropriate probability. And in the converse case, $Q'(z')$ would be appropriate.

Area Maps: The Blocks Test and the Runs Test

The preceding discussion dealt only with "linear maps," in which examined locations could be treated as lying (more or less) in a row. Now consider "area maps," with locations scattered over a fairly isodiametric two-dimensional space (see Figure 9.10). As before, the map shows presences (●) and absences (○) of the species of interest.

It is convenient to discuss the blocks test first, and the runs test afterward.

To perform the blocks test, cuts (partitions) are drawn through the presences in a way that divides the whole area into nonoverlapping, statistically equivalent blocks. We then observe how the absences are distributed among the blocks. As before, the number of empty blocks is the test criterion, and we judge whether or not the observed number is improbably high given that all blocks were equally likely to contain each absence. The theory is identical with that already discussed in connection with linear maps.

The only new problem is the drawing of cuts to demarcate the blocks: how should this be done? It would not be impossible to do as before, that is, use as cuts north-south (say) lines through the presences which would

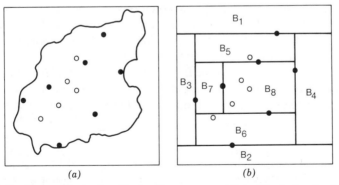

(a)   (b)

Figure 9.10 The partitioning into blocks of an area map. (a) An island, showing the sites of presences (●) and absences (○). (b) The same map, with coastline omitted, partitioned into blocks by cuts through the presences. The cuts were drawn in accordance with the rules given in the text. The labels $B_1, B_2, \ldots,$ mark the first, second, $\ldots,$ blocks thus delimited.

divide the map into $M + 1$ very long narrow blocks. The objection to doing this is that any one such block could contain locations that were very far apart on the ground. What is required is an objective method of constructing more "natural" blocks. There are many ways in which this could be done and a convenient procedure—in effect, a set of instructions to be followed—is outlined below. From a statistical point of view, many other methods are equally good. The essential requirement is that a method be chosen *before* the map to be tested is seen; the chosen instructions must then be carried out impartially. Strict adherence to this rule is the only way of ensuring that biased preconceptions cannot influence the result.

Now consider the map (an imaginary island) in Figure 9.10 a. Presences and absences are shown; numbers have been kept low for clarity. To do a blocks test, cuts were drawn through the $M = 7$ presences. They divided the area into $M + 1 = 8$ blocks.

The cuts were drawn, in the order listed, according to the following rules.* *Step 1:* draw an east-west cut through the northernmost presence and then through the southernmost presence; *step 2:* draw a north-south cut through the westernmost presence and then through the easternmost presence; repeat step 1; repeat step 2; continue thus until a cut has been drawn through every presence. Each cut goes through a presence that has not already been cut, and extends only as far as (but does not cross) a previously drawn cut. For the example in the figure, the result is shown on the right; the island's coast has been omitted for clarity since its irregular shape is of no importance in the test. The blocks are labeled B_1, B_2, . . . , B_8 in the order in which they were formed.

A blocks test can now be performed in the same way as has already been described for a linear map; the equations are unchanged. Thus in the example in Figure 9.10, and using the same symbols as before, $M = 7$, $N = 5$, $z = 5$, and $Q(z) = 456/792 = 0.58$. The converse test could be done by drawing the cuts through the absences.

Figure 9.11 shows an example from real life. The organism whose range was under investigation is the aphid *Cachryphora serotinae,* one of several aphid species that infests goldenrod plants (*Solidago* species). The map shows sites in two eastern provinces of Canada (Nova Scotia and

*In practice, if the sites (presences and absences) are very unevenly spaced, it helps to make a stylized version of the map before drawing the cuts. This is done by assigning a pair of coordinates, (x, y), to every site, with x = rank of the site in a west-to-east listing of the sites and y = rank of the site in a south-to-north listing. The sites can then be plotted on graph paper using these (x, y) coordinates. If a blocks test is done with blocks constructed as described, the results are the same regardless of whether the original map or the stylized version is used (Pielou, 1974).

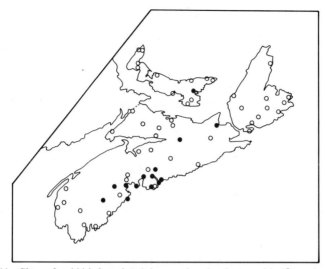

Figure 9.11   Sites of aphid-infested *Solidago puberula* plants, with (●) and without (○) individuals of *Cachryphora serotinae* among the infesting aphids. A blocks test with cuts through the absences gave Q' $(z') = 0.007$. (From Pielou, 1974.)

Prince Edward Island) at which specimens of *Solidago puberula* infested with aphids of any kind were examined; the sites are treated as presences or absences depending on whether the infesting aphids did or did not include *C. serotinae*. A blocks test, performed by drawing cuts through the absences gave $Q'(z') = 0.007$; this led to rejection of the hypothesis that, in the context, seemed the reasonable one to test, namely that *C. serotinae* occurred at random on its *S. puberula* hosts.

Now consider how one might perform a "symmetrical" test on an area map, that is, a test that (unlike a blocks test) treats presences and absences equally. On an area map, there are no obvious "runs" of sites of the two kinds as there are on a linear map. What is formally a runs test can easily be done, however, by first drawing cuts through presences and counting z, the number of blocks empty of absences; and then drawing cuts through absences and counting z', the number of blocks empty of presences. Now treat $k_a = M + 1 - z$ as the number of "runs" of absences and $k_p = N + 1 - z'$ as the number of "runs" of presences. The total number of "runs" is $k_a + k_p = r$. (Equivalently, k_a and k_p are the numbers of nonempty blocks in the respective partitioned maps.)

One can now do a test using r as test criterion in exactly the same way as if the map were linear. The lack of any obvious runs on an area map does not, therefore, preclude the carrying out of a test that treats pres-

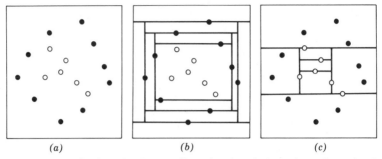

Figure 9.12 Determination of r, the two-dimensional equivalent of one-dimensional runs, from an area map. (a) The unpartitioned map. (b) Cuts through the $M = 10$ presences give $k_a = 1$ nonempty blocks. (c) Cuts through the $N = 6$ absences give $k_p = 4$ nonempty blocks. Thus $r = k_a + k_p = 5$ is the total number of "runs."

ences and absences equally; in other words, one need not, as in the blocks test, treat the two kinds of sites unsymmetrically.

An example is shown in Figure 9.12. Applying the two blocks tests and the runs test to the map in the figure gives, respectively, $Q(z) = 0.0014$ (with cuts through presences); $Q'(z') = 0.5490$ (with cuts through absences); $P(r) = 0.0470$ (for the area "runs" test).

The Spread Test

The spread test is applicable to linear maps and is unsymmetrical. Suppose one observes a total of Y habitable sites arranged more or iess in a row, along a shore or a river valley for example. Suppose the species of interest is present at A of the sites and absent from the remaining $Y - A$. All sites appear equally suitable for the species, but a preponderance of presences in the middle of the row leads to the supposition that the species is not uniformly spread through the area sampled.

We wish to test the null hypothesis that the species was equally likely to be found at any of the examined sites. This may be done by observing the "span" of the A presences, say s; s is defined as the number of sites in the row between and including the first and last presences (equivalently, between and including the two terminal presences). An example is shown in Figure 9.13 in which $Y = 20$, $A = 9$, and $s = 11$.

We now find $\pi(s)$, the probability that the span of the A presences is s, given that they are a random subset of the Y sites. This probability may be found as follows. There are $Y + 1 - s$ ways of choosing a single run of the sites so that their span is s. For each of these ways there are

$$\binom{s - 2}{A - 2}$$

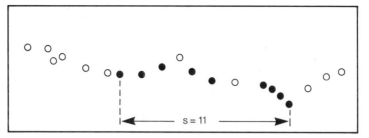

Figure 9.13 The method of counting the span, s, of the presences, in a row of presences (●) and absences (○); s is the test criterion for the spread test.

ways of placing the $A - 2$ nonterminal presences in the $s - 2$ sites between the two terminal presences. The total number of ways of choosing A of the Y sites to be presences, without restriction, is

$$\binom{Y}{A}.$$

Hence

$$\pi(s) = (Y + 1 - s) \frac{\binom{s - 2}{A - 2}}{\binom{Y}{A}}.$$

To test the null hypothesis against the alternative that the observed span of presences is improbably short, we require $\Pi(s) = \Sigma_{j=A}^{s} \pi(j)$; (inevitably $s \geq A$). If this probability is significantly low, then there is reason to reject the null hypothesis and to conclude that the species' range is less extensive than the total sample area. If this area is itself only part of the species' known range, it then follows that the range as a whole is fragmented into disjunct subranges. Some examples, of aphid species with disjunct subranges, have been described by Pielou (1973).

Chapter Ten

THE GEOGRAPHY OF GENES AND CHROMOSOMES

This chapter, like the earlier ones, deals with a few selected topics within a very large subject. One of the most basic questions a biogeographer could ask a geneticist—are there consistent geographic trends in the genetic variability of populations?—has yet to be answered satisfactorily. Speculations about what "should" be the case have been refuted by accounts of what actually is observed; (this shows, if nothing else, the futility of unsupported speculation). What little evidence there is suggests that in at least some taxa, genetic variability decreases with increasing latitude if, as a measure of variability, one takes the percentage of gene loci that are polymorphic (Gould, 1976). (A locus is polymorphic if the commonest allele at the locus has a frequency appreciably less than 100%; how much less is arbitrary; 99% and 95% are values often chosen.) But in a different sense, genetic variability appears to increase with increasing latitude: there is evidence of geographic variation in the frequency of polyploidy in different groups, and the tendency is for percentage polyploidy to increase with increasing latitude.

Visible, phenotypic variability in a population cannot be taken to indicate a matching genotypic variability. It may do so; but it is quite possible for individuals with closely similar genotypes to have highly variable phenotypes if natural selection has favored genes conferring high plasticity. To distinguish the physiological plasticity of individuals from the genetic variability of populations requires time-consuming experiments and this means that conclusions of wide generality on geographic trends in genetic variability will be slow in coming.

Studies on whole genomes can be done more rapidly, however, and there is a greater body of knowledge on geographic variability, within certain taxa, of the number of genomes per somatic cell. This is the subject of Sections 1 and 2. Section 1 describes a few examples of organisms with alternating generations in which the geographic ranges of the diploid and haploid generations do not coincide. Section 2 describes something of what is known about geographic variation in the incidence of polyploidy.

Section 3, a mathematical section, discusses geographic clines and provides an elementary introduction to modern theoretical studies of what happens on an environmental gradient when gene flow is counteracted by natural selection.

1. DIFFERING RANGES IN HAPLOID AND DIPLOID GENERATIONS

Several cases are known in which the haploid (gametophyte) and diploid (sporophyte) generation of plants with alternating generations have non-coincident ranges. The best-known examples are among the marine algae and the ferns.

According to Dixon (1963), there are many species of European Rhodophyta (red algae) in which tetrasporophytic (presumably diploid) plants are found farther north than sexual (presumably haploid) plants. Completely sterile individuals, which produce neither gametes nor asexual spores, are found still farther north in populations that must be maintained either by continual immigration from the south of spores or detached vegetative fragments, or by vegetative reproduction occurring locally. In marine plants, unlike terrestrial ones, obligate vegetative reproduction is no bar to long-distance jump dispersal.

A member of the Phaeophyta with a curious geographic range is *Ectocarpus siliculosus*. It is a small filamentous brown alga with an isomorphic alternation of generations. The alternation is not obligate, however, since each generation can reproduce itself: the diploid plants, besides forming haploid zoospores (in unilocular sporangia), also form diploid zoospores (in plurilocular sporangia) and thus an uninterrupted sequence of diploid plants can result. Likewise the gametes released by the haploid plants, besides fusing in pairs to form zygotes from which diploid plants grow, can also develop parthenogenetically and thus reproduce haploid plants over indefinitely many generations.

It has been found (Knenfuss, 1935; Müller, 1964; and see Chapman and Chapman, 1973) that the two generations are present in different proportions in different geographic regions. In cool northern waters, off the coast of Sweden, the great majority of plants, perhaps all, are diploid. In warm Mediterranean waters, in the Bay of Naples, the opposite is true: nearly all plants are haploid. It is possible that the two generations are adapted to different conditions of light and temperature, though how this can have arisen, and why it should have, are unclear. However, off Woods Hole, Massachusetts, both generations are present, but a small-scale (microgeographic) spatial segregation occurs because the two generations grow as epiphytes on different hosts. The haploid generation appears to be an obligate epiphyte on the larger brown alga *Chordaria flagelliformis* and cannot grow in its absence. The diploid generation occurs more widely since it is less restricted in its choice of hosts; these include (besides *Chordaria*) another brown alga, *Chorda filum,* and the saltmarsh angiosperm *Spartina* (cord grass).

It is not inconceivable that the two generations of a species with a non-obligate alternation of generations could diverge genetically from each other. Bernatowicz (1958) has argued that somatic mutations could appear, in either the diploid or haploid populations, and be perpetuated in the clones of either kind. He regards this process as a probable explanation for the fact that the green alga genus *Derbesia* has more sporophyte than gametophyte "species."

Among terrestrial cryptogams, some ferns are known in which sterile haploid prothalli (sterile "gametophytes") perpetuate themselves without producing diploid sporophytes in areas where the sporophyte generation could not survive because the winters are too cold for them. Farrar (1967) has described this phenomenon in four fern genera, most of whose members are tropical rain forest species. In some species, however, the haploid generation has been found without the diploid; the haploids occur in warm, moist, temperate forests in the Appalachian region of the southeastern United States. Vegetative reproduction is unusual in the haploid generation of ferns; in the geographically isolated haploids in which it is known, it takes place by the formation, on the margin of the prothallus, of very small (10 cells or fewer) dispersable gemmae. The gemmae are spread by air currents and give rise to extensive clones of prothalli which somewhat resemble patches of thallose liverworts. An example of a fern of this type is *Vittaria lineata* (shoestring fern), described by Wagner and Sharp (1963; and see Wagner, 1972). The Appalachian population of this species consists exclusively of sexually nonfunctioning, vegetatively reproducing prothalli. This population is separated, by a disjunction hundreds of kilometers wide, from the normal, ancestral population in which a dominant sporophyte generation alternates with a functioning gametophyte generation.

It is interesting to observe that in these ferns, unlike the marine algae mentioned earlier, it is the haploid generation whose range extends farther, and into a colder environment.

Turning now to phanerogams, it is not possible of course for the alternating generations to be geographically separated, since the haploid generation is minute, short-lived, and parasitic on the diploid plant at all times; thus it cannot perpetuate itself independently. But mismatched geographic ranges sometimes occur in the two "sexes" of dioecious species. An example (Briggs and Walters, 1969) is the herb *Petasites hybridus* (butterbur) of the family Compositae. In Great Britain, although male plants (strictly, microsporophytes) are found everywhere, female plants (strictly, megasporophytes) are of more limited range, being common only in the northern counties of England. Where females are absent, males reproduce themselves vegetatively by means of rhizomes.

As the examples described in this section show, there are plants in which the different generations of a species, or the separate sexes, have noncoincident geographic ranges. These cases tend to be interesting oddities; they are not all of one kind and are not numerous enough for any general conclusions to be based on them. In each case, it would be worth knowing how the geographic ranges of the contrasted populations—haploid versus diploid, or female versus male—of a single species are changing. Has the population of smaller range suffered a contraction of range, or has the wider-ranging population expanded? Are the ranges as now observed of long standing, or are they continually varying? It remains to be seen whether answers can be found to these questions for any of the species about which they might be asked.

2. THE GEOGRAPHY OF POLYPLOIDY

The importance of polyploidy in plant evolution is known to all botanists. Indeed it is quite possible that all angiosperms are descended from ancestors in which the number of chromosomes in a single set (a genome) was 6 or 7, and hence that newly formed polyploids occurred, perhaps in the far distant past, one or more times in the lineages of nearly all modern plants (Stebbins, 1971). The same is no doubt true of cryptogams, especially ferns, in which polyploidy is common. Polyploidy is very much less common in animals than in plants but is known in some insects, as described below. First, however, we consider plants.

To begin, it is interesting to pursue the biogeographic implications of the following familiar fact: in phanerogams the disseminule is a seed, whereas in cryptogams it is a spore. This contrast has interesting consequences (Klekowski, 1972). When a diploid fern species invades a new region by means of a single spore, no genetic variability is possible in the descendants except as a result of mutations. This is because the founding spore, being haploid, grows into a gametophyte giving genetically identical gametes; self-fertilization must then yield a sporophyte that is homozygous at every locus and that produces spores identical with the founding spore. A population descended from a single spore is therefore a clone and cannot spread beyond the environment to which it is adapted.

Now consider the contrasting situation in a diploid phanerogam. The disseminules are seeds carrying the same number of chromosomes as the plant (a sporophyte) that bears them. If the plant that grows from an isolated immigrant seed is self-fertile (as it would have to be, to reproduce sexually), its F_1 descendents could contain many different genotypes; thus if there are heterozygous loci in n of its chromosomes, the F_1 can contain as many as 3^n distinct genotypes (assuming no crossing over). Therefore the population as a whole will probably be adapted to a wider variety of habitats than any one of its individual members and hence will be capable of expanding its range and consolidating its hold on the newly invaded region.

This argument implies that the successful establishment of a colonizing population by a single disseminule is much less likely for ferns than for seed plants. However, as Klekowski (1972) has shown, the argument does not apply if the ferns are polyploid. To see this, consider a tetraploid fern species. Its spores, its gametophytes, and its gametes are all *dihaploid;* that is, they each contain two of the genomes derived from the tetraploid's original (diploid) ancestor. Assume that when bivalents form at meiosis any two chromosomes from each tetrad can pair with each other. Then cross-fertilization between the gametes of a tetraploid fern species is isomorphic with "self-fertilization" in the sporophyte of a diploid phanerogam, and can yield as much genetic variety, with as much chance of successful range expansion. ("Self-fertilization" is here put in quotation marks since it actually connotes a cross between a microgametophyte and a megagametophyte that have grown, respectively, from a pollen grain, or microspore, and an ovule, or megaspore, borne by the same sporophyte. Note also that cross-fertilization between two fern gametophytes yields only one daughter sporophyte, whereas "self-fertilization" in one seed plant may yield numerous daughter sporophytes.)

This may explain, at least in part, the high chromosome numbers typically found in ferns. Possibly an additional contributory cause is that ferns have been in existence much longer than seed plants and therefore there has been more time for chromosome multiplication. In support of Klekowski's explanation is Wagner's (1972) observation that, in Hawaiian ferns, the incidence of polyploidy is only half as frequent among endemic species as among nonendemics; presumably many of the latter are descended from single immigrant spores, and polyploids have fared better than diploids for the reason outlined above.

Now consider polyploidy in seed plants and how it varies geographically. About half of all angiosperm species are polyploid. The age and probable history of many "polyploid complexes" can to some extent be inferred from the modern geographic ranges of their member populations. A *polyploid complex* consists of an array of interrelated populations of diploids and polyploids that has developed from pairwise hybridizations between related species, followed by chromosome doublings, further out-crossings and back-crossings, further doublings, and so on. Stebbins (1971) recognizes a succession of stages in the development of such a complex from initiation, through youth, to maturity, decline, and a final relict stage. In the initial stage, a polyploid complex contains only a few polyploid species and their ranges are nested within those of their more wide-ranging diploid ancestors. Then the ranges of the polyploids expand while those of the diploids contract; later, the ranges of the diploids dwindle until they no longer overlap each other and hence are no longer able to hybridize; ultimately, when the ancestral diploids have gone extinct, there may remain only one or two high polyploid species with no close relatives still extant. As examples of complexes in successive stages of development from youth to old age, Stebbins describes species groups in the genera *Tragopogon* (in the family Compositae), *Aegilops* (Graminae), *Lotus* (Leguminosae), *Dactylis* (Graminae), *Clarkia* (Onagraceae), and *Brasenia* (Nymphaeaceae). It is thought that most complexes now in the mature stage originated in the Pliocene or Pleistocene.

The modes of reproduction of angiosperms tend to vary with their ploidy level, and consequently tend to vary geographically as well. For example, the many races of *Taraxacum officinale* (dandelion) are a polyploid complex. Among them are diploid races that reproduce sexually, and which are native to central Europe. There are also polyploid races that reproduce apomictically, by the asexual production of embryos (agamospermy); many of these races occur in northern and western Europe and have been introduced into North America.

It is noteworthy that the invaders of disturbed or newly opened habitats which, by definition, are "weeds," are often apomictic polyploids; *Taraxacum* is but one example. Moreover the great majority of weeds are herbaceous rather than woody; and polyploidy and apomixis are much less common in trees than in herbs. The chain of cause and effect in this set of related facts is hard to unravel; it has been discussed by Stebbins (1971) who points out that of the few woody angiosperms that do behave as weeds, two of the most aggressive genera—*Crataegus* (the hawthorns) and *Rubus* (the brambles)—both have very high proportions of apomictic polyploids among their members.

Turning now from particular polyploid complexes to the incidence of polyploidy in whole floras, the trend was well illustrated by Haskell (1952) in a study of the angiosperm floras of several European countries and regions. His results are shown in Figure 10.1. Two conclusions stand out. First, in every region the proportion of polyploids is greater among

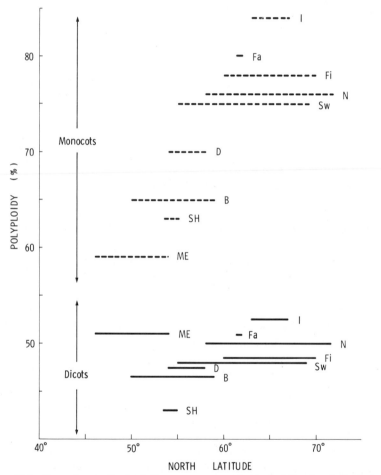

Figure 10.1 Percentages of polyploids among monocots (dashed lines) and dicots (solid lines) in several European countries. The lengths and positions of the lines show the latitudinal limits of the several regions. These are: B, British Isles; D, Denmark; Fi, Finland; Fa, Faeroes; I, Iceland; ME, Mid-Europe; N, Norway; SH, Schleswig-Holstein; Sw, Sweden. (Redrawn from Haskell, 1952.)

monocots than among dicots. Second, in each of the two subphyla, the proportion of polyploids increases with increasing latitude. The large proportion of polyploids among the dicots of the comparatively low-latitude region "Mid-Europe" is the only exception and is due to the high incidence of polyploidy among mountain plants.

Haskell's results demonstrate a relationship that has been found repeatedly in different regions and in many taxonomic groups: the colder the climate, the greater the incidence of polyploidy. Possible reasons are discussed below, but first some additional items of evidence may be mentioned. According to Haskell, one such item consists in the difference between monocots and dicots in their proportions of polyploids as shown in Figure 10.1. He argues that the ancestors of modern terrestrial monocots were aquatic plants and were therefore rooted in cold, waterlogged soil; the roots of these plants were thus always at lower temperatures than those of neighboring land plants, and natural selection may have favored polyploidy in these conditions.

Johnson and Packer (1965) found a relation between polyploidy and soil temperature on an ecological scale that matches the relation with latitude on a geographic scale. These authors investigated the frequency of polyploidy among plants growing in different habitats in a single valley in arctic Alaska. The valley is at 68°N latitude, in a region that was icefree throughout the Pleistocene (see Figure 4.1). Its soils provide a spectrum of conditions. The valley bottom (lowland) soil consists of fine silt and organic material and is always wet and cold; it is subject to intense frost-heaving and the top of the permafrost is only 0.5 m below the surface. The soil of the slopes and ridges (upland) contains a large proportion of coarse rock fragments and is therefore drier and hence warmer than the lowland soil; frost action is comparatively mild and the top of the permafrost is 2 m down. Between these two extremes, Johnson and Packer recognized four intermediate sets of conditions on the basis of recognizably distinct groups of plants; they assigned label numbers to these groups, and to the soils they indicated, as shown in Figure 10.2 which summarizes their results. The relation between frequent polyploidy and cold soils is clear; so also is the greater frequency of polyploidy in monocots than in dicots.

There are two (at least) distinct explanations for the trends shown in Figures 10.1 and 10.2. The first is that polyploids are more likely to be formed in disturbed periglacial environments than elsewhere. The second is that polyploidy is an adaptation to cold and is selectively favored in cold climates. We consider these explanations in turn.

The first has the support of Stebbins (1971) who dismisses as mistaken the notion that polyploidy has evolved in response to climatic cooling. In

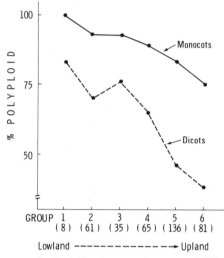

Figure 10.2 Percentages of polyploids among angiosperms of different habitats in an unglaciated Alaskan valley. In parentheses below the plant groups' label numbers are shown the numbers of taxa sampled. (Redrawn from Johnson and Packer, 1965.)

his view, the probable course of events was as follows: with every expansion of the Pleistocene ice sheets, conditions ahead of the advancing ice fronts grew colder and harsher; only the hardiest plant species were able to survive and they were also able to spread into habitats left vacant by species that had succumbed; thus the composition of the vegetation changed, and species populations were brought into proximity in unusual combinations; hybridizations occurred that would have been most unlikely in normal circumstances, and in some cases were followed by chromosome doubling. Further out-crossings, back-crossings, and chromosome doublings would have led to the formation of polyploid complexes with rich gene pools, and from the large assortment of genotypes resulting, many of them polyploid, would have come plants adapted to invade the new habitats that opened up when the ice sheets melted.

This argument would be unlikely to occur to anyone except by hindsight and there seems no compelling reason to rule out the second explanation for frequent polyploidy in high latitudes, namely that polyploids are better adapted physiologically than are diploids to withstand cold. It is true that in the Pleistocene and Holocene, high northern latitudes were (and still are) characterized both by low temperatures and by "disturbed" conditions, the outcome of recent, drastic climatic change, and it is difficult to judge which of these two causes, if either, was paramount in bringing about the high incidence of polyploidy.

The large numbers of polyploids in areas that may have been icefree refugia lend support to Stebbins's view. For example, Nannfeldt (1963) has described the polyploid complex of species and subspecies of the genus *Papaver* (poppy) occurring in the two supposedly icefree "centers" in Scandinavia (see Figure 4.6). Although all members of the genus occurring in the Alps are diploid (with $2n = 14$), the three Scandinavian species are all polyploid; one is octoploid and two are decaploid, and they are found in one or other (or, in the case of *P. radicatum*, in both) of the centers. Indeed, *P. radicatum*, a decaploid species, is of particular interest. In each center it has at least four local races of exceedingly restricted distribution and these races, while internally homogeneous, differ markedly from each other in chromosome structure although in outward appearance they are very similar. These facts, although they relate to only one species, appear to suggest that periglacial conditions have a pronounced effect at least on plants' karyotypes.

It is now interesting to consider evidence that cold itself, regardless of habitat disturbance, may favor polyploidy. It comes from experiments performed by Hall (1972) who compared the oxygen requirements of the root meristems in seedlings of diploid and tetraploid rye (*Secale cereale*). He concluded that tetraploids are less able than diploids to endure high temperatures. The reason is as follows. The oxygen needed for respiration in the internal tissues of a root must diffuse into the root from the soil atmosphere. If the root is thick, diffusion may be so slow that the oxygen concentration near the root axis is insufficient for the cells to respire normally. Thus in warm climates, where soils are warm and respiration rates high, thick roots are a disadvantage; only in thin roots is the oxygen concentration adequate for normal respiration. In cold climates respiration rates are low enough for a thick root not to be at an appreciable disadvantage. Since all plant organs tend to be larger, and roots thicker, in polyploids than in their ancestral diploids, polyploids are less well adapted than diploids to warm soils. Notice that the conclusion is not that polyploids are well adapted to cold but rather that they are ill adapted to warmth.

The tendency for ploidy levels to be higher in high latitudes is not universal. As a counterexample, consider *Epilobium angustifolium* (rosebay willowherb), which has been studied by Mosquin (1967). The species has two principal subspecies. One of them, *angustifolium*, with $2n = 36$ has a circumpolar range at high latitudes. The other, *circumvagum*, with $2n = 72$, occurs farther south in warmer climates. Mosquin comments that "polyploids often have morphological features (large leaves, large stomates, slower growth, etc.) that are potentially adaptive in [warm] climates." It is not clear why, if the polyploids of *Epilobium* are adapted to warmth, the same is not true of polyploids in other genera. Nor do these conclusions accord with Hall's (1972) subsequent findings in his experiments with rye seedlings.

As has been argued above, if diploids and polyploids preponderate in different geographic regions, the implication may be either that the formation of polyploids is more probable in some regions than in others, or that polyploids appear independently of locality but have different survival rates in different environments. It would be interesting to know whether there are geographic or environmental differences in the rates at which polyploids revert to the diploid state. Such reversals are possible but are probably uncommon (Stebbins, 1971). In tetraploids, reversals result from the parthenogenetic development of a gamete (de Wet, 1971). An unfertilized gamete is only likely to develop into a robust individual if tetraploidy in the ancestral lineage is sufficiently recent for most of the gene loci to be still in the tetrasomic condition, that is, with four alleles of a particular gene in every cell of the plant; then the gametes will be dihaploid, that is, they will have every gene locus in duplicate. As a tetraploid lineage ages, its gene loci become converted from the tetrasomic to the disomic condition and its gametes then behave as haploid cells; then parthenogenetic development, if it occurs at all, produces only weak or sterile plants. Natural selection acts most efficiently on diploids. Indeed, in polyploid complexes it is usually the diploid members that exhibit the most extreme forms, both morphologically and ecologically (Stebbins, 1971). Therefore the reversion of a recently formed tetraploid lineage to the diploid state permits further evolution to be more radical and more rapid. The possibility that the reversion rate is affected by climate is worth investigating.

Suomalainen (1961, 1962) has discussed the phenomena of polyploidy and parthenogenesis* in insects. Many insect species have parthenogenetically reproducing as well as bisexual forms, and usually the parthenogenetic form is polyploid and has a wider geographic distribution.

Two examples described by Suomalainen, both from Europe, are the bagworm moth *Solenobia triquetrella* (Lepidoptera: Psychidae) and the weevil *Otiorrhynchus dubius* (Coleoptera: Curculionidae). The bisexual, diploid form of *S. triquetrella* is restricted to areas north of the Alps that are known to have been icefree during the last glaciation (the Würm) and

---

*The type of parthenogenesis considered here is *apomictic parthenogenesis* in which generation after generation of females appear with no males. It is especially common in aphids. Genotypically new forms result only from new mutations (including changes in chromosome structure).

to areas within the Alps that were probably nunataks. The parthenogenetic, tetraploid form is found throughout Europe. The pattern is similar in *O. dubius*: bisexual, diploid individuals are found in mountainous regions in central Europe that were unglaciated during the Würm, and the parthenogenetic, tetraploid form is widespread in northern Europe.

The larger geographic ranges of the parthenogenetic forms may be due to the greater ease with which they can spread to colonize new areas; a single female can establish a new colony. Moreover polyploid forms, if they are polysomic (with each gene locus represented by more than two replicates) are more likely to enjoy the benefits of heterozygosity, in particular to have wider ecological amplitudes, and this should enable them to occupy more extensive ranges. Thus parthenogenesis and polyploidy are both conducive of large geographic ranges, though by entirely different mechanisms. It is difficult to disentangle the relative importance of these two causes.

*3. GENE FLOW ALONG A GEOGRAPHIC GRADIENT

This section considers gene flow on a geographic scale. The problem to be investigated is this: what happens when gene flow and natural selection tend to counteract each other? Theoretical studies of this question have a long history. The work to be reviewed here is due especially to Haldane (1948), Fisher (1950), Slatkin (1973), and May, Endler, and McMurtrie (1975). The rigorous mathematical derivations in the original papers are of a fairly high level of difficulty; here, a more heuristic, introductory approach is attempted.

Imagine, first, the outcome of the following conceptual experiment. Two interfertile populations of a motile species, distinguished by their different frequencies of a marker gene, are placed on either side of a barrier across an environmentally uniform tract of land. The barrier is suddenly removed. Clearly, the hitherto separated populations will now become mingled as a result of intermigration, cross-breeding, or both. It makes no difference whether one or other of these processes acts alone or whether both act together; the net result is that gene flow occurs and the populations diffuse into each other. With the passage of time, the initial difference between the originally separated populations in their frequency of the marker gene will flatten out.

The two curves in Figure 10.3 show, qualitatively, how the frequency of the marker gene would be expected to change in the time interval from t to $t + 1$ assuming that at $t = 0$ its frequency was given by the step function $y(0)$. Gene frequency, y, is measured on the ordinate; distance from the

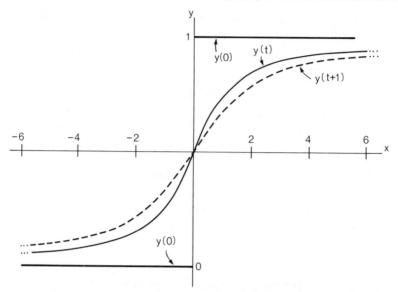

Figure 10.3   The step function $y(0)$ and two curves, $y(t)$ and $y(t + 1)$, for which $y(t + 1) - y(t) = d^2y/dx^2$. See text for explanation. These curves are *NOT* solutions to the diffusion equation. They were constructed by assigning to $y(t)$ an elementary function of the desired S-shape, namely $y(t) = (2/\pi)$ arctan x, for which d^2y/dx^2 has the same sign as x. Then $y(t + 1)$ was found by putting $y(t + 1) = y(t) + d^2y(t)/dx^2$. Thus (over one time unit only) $y(t)$ and $y(t + 1)$ behave in the desired way.

former barrier, x, is measured along the abscissa, with $x = 0$ marking the site where the barrier (at right angles to the page) used to be. It is assumed that the frequency of the gene was initially lower in the "left" population than in the "right" population and that, with the passage of time, the difference between the frequencies to left and right decreases.

Let $y(t)$ and $y(t + 1)$ denote the curves of y versus x at times t and $t + 1$. Put $\delta y = y(t + 1) - y(t)$. We wish to find a simple equation for δy which will ensure that the shapes of the curves $y(t)$ and $y(t + 1)$ shall resemble, qualitatively, those shown in the figure.

Consider $y(t)$. To the left of $x = 0$, $y(t)$ is concave up which implies that $d^2y/dx^2 > 0$; to the right of $x = 0$, $y(t)$ is convex up which implies that $d^2y/dx^2 < 0$; and at $x = 0$, $y(t)$ has an inflection which implies that $d^2y/dx^2 = 0$. Thus for all x, δy has the same sign as d^2y/dx^2. The simplest possible equation describing a change from $y(t)$ to $y(t + 1)$ like that shown in the figure is therefore obtained by letting δy be proportional to d^2y/dx^2. That is, we put

$$\delta y = D \frac{d^2 y}{dx^2} \quad , D > 0 \tag{10.1}$$

where D is a constant of proportionality.

This is the one-dimensional diffusion equation of physics. The foregoing paragraph is intended to show its intuitive reasonableness. More rigorous derivations are found in textbooks of physics [or, with an ecological slant, in Pielou (1977)]. The equation is not explicitly solvable. That is, no elementary function exists that satisfies (10.1).

Hitherto it has been assumed that environmental conditions were the same to left and right of the barrier. Now drop this assumption and suppose, instead, that the barrier marks the position of a sharp change in environment. The "left" area ($x < 0$) and the "right" area ($x > 0$) are still internally uniform, however.

Suppose that the difference between the left and right populations is due to the gene pair **A** and **a**. The homozygote **aa** is selectively favored where $x < 0$ and the homozygote **AA** where $x > 0$. The fitness of the heterozygote **Aa** is intermediate between that of the homozygotes. For concreteness, assume that the relative fitnesses of the three genotypes **aa**, **Aa** and **AA** are in the ratio

$$1 + k : 1 : 1 - k \qquad \text{where } x < 0$$

and

$$1 - k : 1 : 1 + k \qquad \text{where } x > 0;$$

k is small and positive. (Symmetry has been assumed for simplicity.)

Now suppose that the barrier is in place. Until further notice, we consider events to the right of it (where $x > 0$) only. Everywhere in this region let the gene ratios at time t be

$$(1 - y) \, \textbf{a} : y \, \textbf{A} \qquad \text{with } 0 < y < 1.$$

Then the genotype ratios are

$$(1 - y)^2 \, \textbf{aa} : 2y(1 - y) \, \textbf{Aa} : y^2 \, \textbf{AA}.$$

Natural selection operating for one generation will cause these ratios to change to

$$(1 - k)(1 - y)^2 \, \textbf{aa} : 2y(1 - y) \, \textbf{Aa} : (1 + k)y^2 \, \textbf{AA}.$$

Hence the relative frequency of gene **A**, say, changes from y to

$$\frac{y(1 - y) + (1 + k)y^2}{(1 - k)(1 - y)^2 + 2y(1 - y) + (1 + k)y^2}$$

$$= \frac{y + ky^2}{1 + ky^2 - k(1 - y)^2}$$
$$= (y + ky^2)\,[1 + k(2y - 1)]^{-1}$$
$$= (y + ky^2)\,[1 - k(2y - 1) + \text{terms in } k^2, k^3, \text{ etc.}]$$
$$\simeq (y + ky^2)\,[1 - k(2y - 1)]$$
$$\simeq y + ky(1 - y),$$

since k is small enough for terms in k^2 and higher powers to be neglected.

Thus the change, δy, in the relative frequency of gene **A** on the right of the barrier, brought about by natural selection in one generation, is

$$\delta y = ky(1 - y). \tag{10.2}$$

Since $k > 0$ and $0 < y < 1$, it is seen that $\delta y > 0$. Thus where $x > 0$, the relative frequency of gene **A** will increase; with the passage of time, it will approach unity asymptotically, so long as the barrier prevents gene flow.

Finally, imagine the barrier to be removed and gene flow (diffusion) to commence. Diffusion will tend to equalize gene frequencies in the hitherto separated areas, but selection will tend to counteract the effect of diffusion. Therefore equilibrium will be attained when the diffusion and selection rates balance. From (10.1) and (10.2) it is seen that this occurs when

$$D\frac{d^2y}{dx^2} = ky(1 - y). \tag{10.3}$$

The effect of diffusion will obviously be to blur the abrupt discontinuity in gene frequencies that selection without diffusion would bring about. The relative frequency of gene **A** at different distances from the stepwise environmental discontinuity, once equilibrium is established, thus qualitatively resembles either of the curves in Figure 10.3. The relation is described by (10.3) with appropriately chosen constants.

Notice that the relation is the resultant of diffusion and selection acting simultaneously; their effects cannot be separated. Therefore two constants are superfluous and can be replaced by one, say C, with $C = k/D$. Notice also that in the region we are considering, where $x > 0$, the curve of y versus x is convex up; equivalently $d^2y/dx^2 < 0$.

Hence if $C > 0$ (by definition), we must have

$$\frac{d^2y}{dx^2} = -Cy(1 - y) \qquad \text{with } x > 0. \qquad (10.4a)$$

Analogously, to the left of the environmental discontinuity, where the curve is concave up,

$$\frac{d^2y}{dx^2} = Cy(1 - y) \qquad \text{with } x < 0. \qquad (10.4b)$$

The boundary conditions are that $y = 0$ at $x = -\infty$ and $y = 1$ at $x = +\infty$.

Hence the solution of the pair of differential equations (10.4a) and (10.4b), subject to these boundary conditions, gives the equilibrium relation between y, the relative frequency of gene **A**, and x, the distance from the environmental discontinuity.

From symmetry it is seen that the curve for gene **a** is the mirror image of that for gene **A**.

In sum, the pair of equations (10.4) constitutes a model (one of the infinity of possible models) that specifies exactly the way in which the relative frequency of a gene changes across a stepwise environmental discontinuity if it is selectively favored on one side and disfavored on the other. The model specifies exactly the way in which a gene's frequency changes gradually, because of diffusion, even though the environmental change is abrupt.

We now extend the discussion to take account of what happens when the environmental change is not abrupt. Let the change in conditions be linear, and let it take place across a transition zone of width Δ (see Figure 10.4). Suppose the relative fitnesses of the genotypes **aa**, **Aa**, and **AA** respectively are in a ratio dependent on x, namely

$$1 - kf(x) : 1 : 1 + kf(x)$$

where $f(x)$ is given by

$$f(x) = \begin{cases} -1 & \text{where } x < -\dfrac{\Delta}{2}, \\[2mm] \dfrac{2x}{\Delta} & \text{where } -\dfrac{\Delta}{2} < x < +\dfrac{\Delta}{2}, \\[2mm] +1 & \text{where } x > +\dfrac{\Delta}{2}. \end{cases} \qquad (10.5)$$

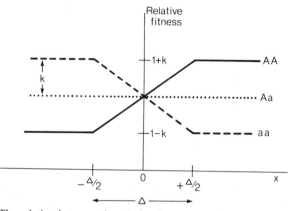

Figure 10.4 The relation between the relative fitnesses of the genotypes **aa**, **Aa**, and **AA** assumed by the model described in the text. The equations are given in (10.5).

Equations (10.4) must now be modified accordingly. As before, put $C = k/D$. Then (10.4) becomes

$$\frac{d^2y}{dx^2} = -C\,f(x)\,y(1-y). \qquad (10.6)$$

Notice that (10.6) simplifies to (10.4) when $\Delta = 0$, that is, when the environmental transition zone is of zero width.

The way in which the solution of (10.6) behaves with different values of Δ, k, and D has been examined by Slatkin (1973) (and see May, Endler, and McMurtrie, 1975). Keep in mind that Δ is the width of the environmental transition zone; that k measures the selective advantage of each homozygote in its own region; and that D is a measure of the rate of diffusion (or gene flow) across the transition zone.

We also require a symbol, say W, for the width of the gene frequency *cline*, that is, the width of the zone within which the changes in gene frequency occur. Since, theoretically, the relative frequencies of **a** and **A** approach unity asymptotically as $x \to -\infty$ and $x \to +\infty$ respectively, there is (in theory) no finite value of x for which the rate of change is exactly zero. Therefore, for convenience, one may take W as the distance between the points at which $y = 0.2$ and $y = 0.8$.

Now observe that, depending on circumstances, we may have $W > \Delta$ or $W < \Delta$. This accords with intuition.

Thus if gene flow is fast and selection weak (large D and small k), gene flow will swamp the effect of selection and the biological cline will be wider than the environmental transition zone; that is, $W > \Delta$. This point has been discussed above for the case $\Delta = 0$.

Conversely, if gene flow is slow and selection strong (small D and large k), selection will be able to "undo" some of the effect of gene flow; each genotype will be able to dominate its own side of the transition zone except in a comparatively narrow strip; that is, $W < \Delta$. In these circumstances the populations on either side of the transition zone can become genetically differentiated in spite of a certain amount of gene flow between them. Then, provided k is large for sufficiently many genes, the conditions are such as to permit parapatric speciation (see Chapter 3, page 83).

Slatkin (1973) has considered the behavior of (10.6) in detail. He found that for a sufficiently narrow transition zone, the width of the genetic cline was independent of the width of the transition zone; it depended only on the rate of gene flow and the strength of selection, that is, on biological as opposed to environmental, factors. Only if the width of the transition zone exceeds a critical value, say l_c, which is a function of k and D, will the width of the cline vary with the width of the transition zone. The quantitative relation is

$$W \simeq l_c \qquad \text{if } \Delta \ll l_c;$$

and

$$W \simeq (l_c^2 \, \Delta)^{1/3} \qquad \text{if } \Delta \gg l_c.$$

For intermediate values of Δ, $l_c < W < (l_c^2 \, \Delta)^{1/3}$.

Notice that for large Δ (a wide transition zone), the gene frequency cline is *steeper* than the selection gradient. Conversely, for small Δ the opposite is true.

Figure 10.5 illustrates these theoretical results diagrammatically. The salient point is that, given the model, there is a characteristic distance, l_c, such that the population cannot change as fast as the environment if environmental changes occur over a distance less than l_c.

The applicability of the foregoing theoretical studies to real life has been examined by Endler (1977) who summarized a multitude of examples of biological clines described in the literature. Their widths range from 2 m to 2000 km. Considering only clines with $W \geq 10$ km (which is here used as an arbitrary boundary between "geographical" and "ecological" clines), the organisms whose clines he tabulates comprise 8 species of mammals, 4 of birds, 1 reptile, 2 amphibians, 4 fishes (including *Pimephales promelas*; see Figure 7.6), 2 snails, 16 insects (including *Biston betularia;* see below), and 1 marine ectoproct. It is noteworthy that

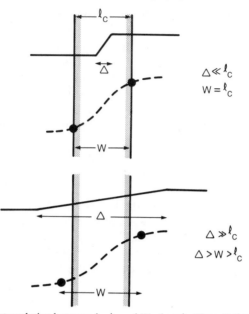

Figure 10.5 The interrelation between l_c, Δ, and W when $\Delta \ll l_c$ and when $\Delta \gg l_c$. The solid lines show $f(x)$ versus x (as in Figure 10.4). The dashed curves show how y, the relative frequency of gene **A**, varies with x, the distance from the midpoint of the transition zone. Dots on the curves show the points at which $y = 0.2$ and 0.8; W is the distance between these points. When $\Delta \ll l_c$, the width of the cline does not depend on the width of the environmental transition zone.

there are no plant species in this list. Judging from Endler's review, clines are fewer and narrower (2 km at most) in phanerogams, and no cryptogams are cited at all. It is unclear to what extent this is merely a reflection of investigators' interests.

Now consider a real example of a cline that is probably maintained by the opposing forces of gene flow and selection. The organism is *Biston betularia* (Lepidoptera; Selidosemidae), the peppered moth. It occurs in Europe in two common forms.* In heavily industrialized regions where nearly all exposed outdoor surfaces are blackened with deposits of soot (or were, when coal was the chief fuel), a melanic form of the moth, *carbonaria*, predominates; its black color camouflages it on sooty surfaces and protects it from bird predators. In rural areas with clean air, the typical rural form predominates; it has a pale pepper-and-salt pattern that camouflages it well when it settles on tree bark and the like.

---

*A third, rather rare form, *insularia*, is not considered here.

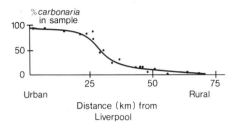

Figure 10.6 A gene frequency cline in the melanic form *carbonaria* of *Biston betularia,* the peppered moth. The cline occurs across a transition stretching from an urban industrial region (Liverpool) to thinly populated countryside. (Redrawn from Bishop, 1972.)

The species thus presents a classic case of industrial melanism and its genetics have been thoroughly investigated (Ford, 1971). It is known that wing color is controlled by a single gene; the melanic form is dominant and the typical form recessive. Studies of bird predation on the moth have shown how effective the protective coloration is; thus the melanics are very conspicuous when they settle on such surfaces as lichen-covered tree trunks in the country and are an easy target for insectivorous birds, whereas the light pepper-and-salt form is almost invisible in the same setting. Conversely, the light form is glaringly obvious in urban surroundings where the melanics are well concealed.

Bishop (1972) investigated the relative proportions of the two forms across a geographical transition zone stretching from the industrial city of Liverpool in northwestern England into rural North Wales. Corresponding with the environmental transition is a clear gene cline in the moths (see Figure 10.6). The proportion of *carbonaria* individuals in samples of *B. betularia* taken from 19 sites across the transition zone was found to range from over 95% at the urban end to 0% at the rural end, 70 km away. The width, W, of the cline, that is, the distance over which the percentage of melanics changed from 20% to 80%, is 15 km (Endler, 1977).

Bishop (1972) tested various diffusion-versus-selection models that might have yielded a predicted cline comparable with the observed one, but there were marked discrepancies between observation and expectation. The discrepancies suggest that selection against the melanics in rural areas is not as strong as had been supposed.

The failure of the theoretical models to match the empirical observations should not be a cause for disappointment since a model that does *not* fit real data is nearly always more informative than one that does. A "good" fit is often more apparent than real, whereas a bad fit usually permits the rejection of a false hypothesis and to that extent constitutes a gain in knowledge. It is by considering a great multitude of concepts, the majority of which will turn out to be mistaken when tested, that biogeography will most surely advance.

BIBLIOGRAPHY AND AUTHOR INDEX

(The italic numbers in parentheses at the end of each reference refer to the pages on which each publication is mentioned)

Abbott, I. (1974). Numbers of plant, insect and land bird species on nineteen remote islands in the southern hemisphere. *Biol. J. Linn. Soc.* **6**: 143–152. *(183)*

Abbott, I. and P. R. Grant (1976). Nonequilibrial bird faunas on islands. *Amer. Natur.* **110**: 507–528. *(183)*

Abbott, R. T. (1968). *Seashells of North America*. Golden Press, New York. *(161)*

Ager, D. V. (1963). *Principles of Paleoecology*. McGraw-Hill, New York. *(231)*

Antevs, E. (1929). Maps of the Pleistocene glaciations. *Bull. Geol. Soc. Amer.* **40**: 631–720. *(109)*

Arber, E. A. N. (1905). Catalogue of the fossil plants of the *Glossopteris* Flora. British Museum (Natural History), London. *(46)*

Ashlock, P. D. (1974). The uses of cladistics. *Ann. Rev. Ecol. Syst.* **5**: 81–99. *(68n,78)*

Ashlock, P. D. and G. G. E. Scudder (1966). A revision of the genus *Neocrompus* China (Hemiptera-Heteroptera: Lygaeidae). *Pacific Insects* **8**: 686–694. *(78)*

Axelrod, D. I. (1970). Mesozoic paleogeography and early angiosperm history. *Bot. Rev.* **36**: 277–319. *(38,43,48,270)*

Axelrod, D. I. (1972). Ocean-floor spreading in relation to ecosystem problems. In *A Symposium on Ecosystematics* (R. T. Allen and F. C. James, Eds.). University of Arkansas, Fayetteville. *(38,43,198,270)*

Axelrod, D. I. and P. H. Raven (1978). Late Cretaceous and Tertiary Vegetation History of Africa. In *Biogeography and Ecology of Southern Africa* (M. J. A. Werger, Ed.). Junk, The Hague. *(43)*

Baker, H. G. (1959). The contribution of autecological and genecological studies to our knowledge of the past migrations of plants. *Amer. Natur.* **93:** 255–272. *(277)*

Ball, I. R. (1975). Nature and formulation of biogeographical hypotheses. *Syst. Zool.* **24:** 407–430. *(61)*

Bé, A. W. H. and D. S. Tolderlund (1971). Distribution and ecology of living planktonic Foraminifera in surface waters of the Atlantic and Indian Oceans. In *Micropalaeontology of Oceans* (B. M. Funnell and W. R. Riedel, Eds.). Cambridge University Press, Cambridge. *(157)*

Beals, E. W. (1969). Vegetational change along altitudinal gradients. *Science* **165:** 981–985. *(228)*

Beck, C. B. (1976). Origin and early evolution of Angiosperms: a perspective. In *Origin and Early Evolution of Angiosperms* (C. B. Beck, Ed.). Columbia University Press, New York. *(38)*

Benson, R. H. (1975). The origin of the psychrosphere as recorded in changes of deep-sea ostracode assemblages. *Lethaia* **8:** 69–83. *(158)*

Berger, W. H. (1969). Ecologic patterns of living planktonic Foraminifera. *Deep-Sea Res.* **16:** 1–24. *(144)*

Bernatowicz, A. J. (1958). Ecological isolation and independent speciation of the alternate generations of plants. *Biol. Bull. Woods Hole* **115:** 323. *(296)*

Bishop, J. A. (1972). An experimental study of the clines of industrial melanism in *Biston betularia* (L.) (Lepidoptera) between urban Liverpool and rural North Wales. *J. Anim. Ecol.* **41:** 209–243. *(313)*

Böcher, T. W. (1963). Phytogeography of Greenland in the light of recent investigations. In *North Atlantic Biota and Their History* (A. Löve and D. Löve, Eds.). Pergamon Press, New York. *(254)*

Bock, W. J. (1969). Review of Henning's "Phylogenetic Systematics" and Brundin's "Transantarctic relationships and their significance." *Syst. Zool.* **18:** 105–115. *(66n)*

Bolli, H. M. (1971). The direction of coiling in planktonic Foraminifera. In *The Micropalaeontology of Oceans* (B. M. Funnell and W. R. Riedel, Eds.). Cambridge University Press, Cambridge. *(158)*

Boltovskoy, E. (1969). Foraminifera as hydrological indicators. In *Proc. First International Conference on Planktonic Microfossils* (P. Brönniman and H. H. Renz, Eds.). E. J. Brill, Leiden. *(155)*

Bourne, W. R. P. (1963). A review of oceanic studies of the biology of sea birds. *Proc. 13th Intern. Ornithological Congr.* **2:** 831–854. *(210)*

Bousfield, E. L. and M. L. H. Thomas (1975). Postglacial changes in distribution of littoral marine invertebrates in the Canadian Atlantic region. In *Environmental Change in the Maritimes* (J. G. Ogden III and M. J. Harvey, Eds.). Nova Scotian Inst. of Science, Halifax. *(114,271)*

Briggs, D. W. and S. M. Walters (1969). *Plant Variation and Evolution.* McGraw-Hill, New York. *(277n,297)*

Briggs, J. C. (1974). *Marine Zoogeography.* McGraw-Hill, New York. *(140,151,164,169,251)*

Brown, J. H. and A. Kodric-Brown (1977). Turnover rates in insular biogeography: effect of immigration on extinction. *Ecology* **58:** 445–449. *(177)*

Brundin, L. (1967). Insects and the problem of austral disjunctive distribution. *Ann. Rev. Ent.* **12:** 149–168. *(63,72,76,79n)*

Brundin, L. (1972a). Evolution, causal biology and classification. *Zool. Scripta* **1:** 107–120. *(72,76)*

Brundin, L. (1972b). Phylogenetics and biogeography. *Syst. Zool.* **21:** 69–79. *(66)*

Burt, W. H. (1958). The history and affinities of recent land mammals of western North America. In *Zoogeography* (C. L. Hubbs, Ed.). AAAS, Washington, D.C. *(271)*

Burt, W. H. and R. P. Grossenheider (1961). *A Field Guide to the Mammals.* Houghton Mifflin, Boston. *(9)*

Bush, G. L. (1975). Modes of animal speciation. *Ann. Rev. Ecol. Syst.* **6:** 339–364. *(82)*

Byers, G. W. (1969). Review of Hennig's "Phylogenetic Systematics" and Brundin's "Transantarctic relationships and their significance" *Syst. Zool.* **18:** 105–107. *(73,76)*

Carlquist, S. (1974). *Island Biology.* Columbia University Press, New York. *(246n,256,259,265)*

Chapman, V. J. and D. J. Chapman (1973). *The Algae.* 2nd ed., Macmillan, London. *(295)*

Charig, A. J. (1973). Kurtén's theory of ordinal variety and the number of the continents. In *Implications of Continental Drift to the Earth*

Sciences, Vol. 1. (D. H. Tarling and S. K. Runcorn, Eds.). Academic Press, New York. *(33,56n)*

Cifelli, R. (1969). Radiation of Cenozoic planktonic Foraminifera. *Syst. Zool.* **18:** 154–168. *(156,159)*

Colbert, E. H. (1971). Tetrapods and continents. *Quart. Rev. Biol.* **46:** 250–269. *(50,51)*

Colinvaux, P. A. (1967). Quaternary vegetational history of arctic Alaska. In *The Bering Land Bridge* (D. M. Hopkins, Ed.). Stanford University Press, Stanford. *(131,135)*

Connell, J. H. (1961). The influence of interspecific competition and other factors on the distribution of the barnacle *Chthamalus stellatus. Ecology* **43:** 710–723. *(227)*

Constance, L. (1963). Amphitropical relationships in the herbaceous flora of the Pacific coast of North and South America: A symposium. Introduction and historical review. *Quart. Rev. Biol.* **38:** 109–116. *(271,273)*

Cormack, R. M. (1971). A review of classification. *J. Roy. Statist. Soc.* **A134:** 321–367. *(17,24)*

Cox, C. B. (1973). Systematics and plate tectonics in the spread of marsupials. In *Organisms and Continents through Time* (N. F. Hughes, Ed.). Paleontological Assn., Special Papers in Paleontology, No. 12. *(54)*

Cox, C. B. (1974). Vertebrate paleodistributional patterns and continental drift. *J. Biogeography* **1:** 75–94. *(32,249)*

Cracraft, J. (1973). Continental drift, paleoclimatology, and the evolution and biogeography of birds. *J. Zool. London* **169:**455–545. *(52,56)*

Cracraft, J. (1974). Phylogenetic models and classification. *Syst. Zool.* **23:** 71–90. *(66)*

Crawford, A. R. (1974). A greater Gondwanaland. *Science* **184:** 1179–1181. *(34,51)*

Croizat, L., G. J. Nelson, and D. E. Rosen (1974). Centers of origin and related concepts. *Syst. Zool.* **23:** 265–287. *(79)*

Cushing, E. J. (1965). Problems in the Quaternary phytogeography of the Great Lakes region. In *The Quaternary of the United States* (H. E. Wright Jr. and D. G. Frey, Eds.). Princeton University Press, Princeton. *(113)*

Damman, A. W. H. (1965). The distribution pattern of northern and southern elements in the flora of Newfoundland. *Rhodora* **67:** 363–392. *(123)*

Dansereau, P. (1957). *Biogeography: An Ecological Perspective*. Ronald Press, New York. *(204)*

Darlington, P. J. (1957). *Zoogeography: The Geographical Distribution of Animals*. Wiley, New York. *(9,12,54,56,59,192,245,251)*

Darlington, P. J. (1970). A practical criticism of Hennig-Brundin "phylogenetic systematics" and Antarctic biogeography. *Syst. Zool.* **19:** 1–18. *(76)*

de Wet, J. M. J. (1971). Reversible tetraploidy as an evolutionary mechanism. *Evolution* **25:** 545–548. *(304)*

Diamond, J. M. (1972). Biogeographic kinetics: estimation of relaxation times for avifaunas of southwest Pacific islands. *Proc. Nat. Acad. Sci. U.S.A.* **69:** 3199–3203. *(186)*

Diamond, J. M. and R. M. May (1976). Island biogeography and the design of natural reserves. In *Theoretical Ecology. Principles and Applications* (R. M. May, Ed.). W. B. Saunders Co., Philadelphia. *(133,197,264)*

Dietz, R. S. and J. C. Holden. (1970). The break-up of Pangaea. *Sci. American* **223(4):** 30–41. *(27,28,37,153)*

Dixon, P. S. (1963). Variation and speciation in marine Rhodophyta. In *Speciation in the Sea* (J. P. Harding and N. Tebble, Eds.). The Systematics Association, London, Publication No. 5. *(295)*

Dodson, M. M. and A. Hallam (1977). Allopatric speciation and the fold catastrophe. *Amer. Natur.* **111:** 415–433. *(85)*

Drury, W. H. Jr. (1969). Plant persistence in the Gulf of St. Lawrence. In *Essays in Plant Biogeography and Ecology* (K. N. H. Greenidge, Ed.). Nova Scotia Museum, Halifax. *(120)*

Durazzi, J. T. and F. G. Stehli (1972). Average generic age, the planetary temperature gradient, and pole location. *Syst. Zool.* **21:** 384–389. *(163)*

Edmunds, G. F. (1972). Biogeography and evolution of Ephemeroptera. *Ann. Rev. Ent.* **17:** 21–42. *(76)*

Ehrlich, P. R. and P. H. Raven (1969). Differentiation of populations. *Science* **165:** 1228–1232. *(84)*

Ehrlich, P. R., A. H. Ehrlich, and J. P. Holdren (1977). *Ecoscience: Population, Resources, Environment*. Freeman, San Francisco. *(139)*

Einarsson, T. (1963). Some chapters of the Tertiary history of Iceland. In *North Atlantic Biotas and Their History* (A. Löve and D. Löve, Eds.). Pergamon Press, New York. *(251)*

Ekman, S. (1953). *Zoogeography of the Sea*. Sidgwick and Jackson, London. *(140,169)*

Eldredge, N. (1971). The allopatric model and phylogeny in Paleozoic invertebrates. *Evolution* **25**: 156–167. *(88)*

Eldredge, N. and S. J. Gould (1972). Punctuated equilibria: an alternative to phyletic gradualism. In *Models in Paleobiology* (T. J. M. Schopf, Ed.). Freeman, Cooper and Co., San Francisco. *(88)*

Eldredge, N. and S. J. Gould (1976). Rates of evolution revisited. *Paleobiology* **2**: 174–179. *(89)*

Elson, J. A. (1969). Late Quaternary marine submergence of Quebec. *Rev. geogr. Montreal* **23**: 247–258. *(113,114)*

Elton, C. (1927). *Animal Ecology*. Tenth Impression, 1966, October House, New York. *(94n)*

Endler, J. A. (1977). *Geographic Variation, Speciation and Clines*. Princeton University Press, Princeton. *(81,83,311,313)*

Ericson, D. B. and G. Wollin (1962). Micropaleontology. *Sci. Amer.* **207(1)**: 97–106. *(144)*

Ericson, D. B. and G. Wollin (1964). *The Deep and the Past*. Knopf, New York. *(144)*

Fairbridge, R. W. (1973). Glaciation and plate migration. In *Implications of Continental Drift to the Earth Sciences*, Vol. 1 (D. H. Tarling and S. K. Runcorn, Eds.). Academic Press, New York. *(111)*

Farrar, D. R. (1967). Gametophytes of four tropical fern genera reproducing independently of their sporophytes in the southern Appalachians. *Science* **155**: 1266–1267. *(296)*

Feller, W. (1968). *An Introduction to Probability Theory and its Applications*, Vol. 1, 3rd ed. Wiley, New York. *(283)*

Ferris, V. R., C. G. Goseco, and J. M. Ferris (1976). Biogeography of free-living soil nematodes from the perspective of plate tectonics. *Science* **193**: 508–510. *(61)*

Fischer, A. G. (1960). Latitudinal variations in organic diversity. *Evolution* **14**: 64–81. *(167)*

Fisher, R. A. (1950). Gene frequencies in a cline determined by selection and diffusion. *Biometrics* **6**: 353–361. *(305)*

Flerow, C. C. (1967). On the origin of the mammalian fauna of Canada. In *The Bering Land Bridge* (D. M. Hopkins, Ed). Stanford University Press, Stanford. *(131,135,136)*

Flint, R. F. (1957). *Glacial and Pleistocene Geology*. Wiley, New York. *(127,216)*

Ford, E. B. (1965). *Genetic Polymorphism.* Faber and Faber, London. *(313)*

Fosberg, F. R. (1963). Plant dispersal in the Pacific. In *Pacific Basin Biogeography* (J. L. Gressitt, Ed.). Bishop Museum Press, Honolulu. *(259)*

Frenkel, R. E. and C. M. Harrison (1974). An assessment of the usefulness of phytosociological and numerical classificatory methods for the community biogeographer. *J. Biogeography* **1:** 27–56. *(25)*

Fuchs, V. (1960). In "Discussion" in *Proc. Roy. Soc. Lond. B.* **152:** 640. *(244)*

Funnell, B. M. (1971). Post-Cretaceous biogeography of oceans—with especial reference to plankton. In *Faunal Provinces in Space and Time* (F. A. Middlemiss and P. F. Rawson, Eds.). Seel House Press, Liverpool. *(141,143,153,154,156,158)*

Gabriel, K. R. (1966). Simultaneous test procedures for multiple comparisons on categorical data. *J. Amer. Statist. Ass.* **61:** 1081–1096. *(237)*

Gabriel, K. R. and R. R. Sokal (1969). A new statistical approach to geographic variation analysis. *Syst. Zool.* **18:** 259–278. *(234)*

George, W. (1964). *Animal Geography.* Heinemann, London. *(8)*

Gillespie, R. et al. (1978). Lacefield swamp and the extinction of the Australian megafauna. *Science* **200:** 1044–1048. *(129)*

Gilpin, M. E. and J. M. Diamond (1976). Calculation of immigration and extinction curves from the species-area-distance relation. *Proc. Nat. Acad. Sci. U.S.A.* **73:** 4130–4134. *(187,190)*

Gjaerevoll, O. (1963). Survival of plants on nunataks in Norway during the Pleistocene glaciation. In *North Atlantic Biota and Their History* (A. Löve and D. Löve, Eds.). Pergamon Press, New York. *(118)*

Gnanadesikan, R., J. R. Kettering, and J. M. Landwehr (1979). Interpreting and assessing the results of cluster analyses. *Proc. 41st. Congr. Intern. Statistical Institute,* Delhi. *(25)*

Godley, E. J. (1967). Widely distributed species, land bridges and continental drift. *Nature* **214:** 74–75. *(262,276)*

Good, R. (1974). *The Geography of the Flowering Plants.* 4th ed. Longmans, London. *(49,74,193,199,219,223,249,268)*

Goodall, D. W. (1973). Sample similarity and species correlation. In *Handbook of Vegetation Science* (R. H. Whittaker, Ed.). Junk, The Hague. *(24)*

Gould, S. J. (1976). Paleontology plus ecology as paleobiology. In

Theoretical Ecology. Principles and Applications (R. M. May, Ed.). W. B. Saunders Co., Philadelphia. *(94,294)*

Greenslade, P. J. M. (1968). Island patterns in the Solomon Islands bird fauna. *Evolution* **22**: 751–761. *(195)*

Greenslade, P. J. M. (1969). Insect distribution patterns in the Solomon Islands. *Phil. Trans. Roy. Soc. B.* **255**: 271–284. *(195)*

Gressitt, J. L. and C. M. Yoshimoto (1963). Dispersal of animals in the Pacific. In *Pacific Basin Biogeography* (J. L. Gressitt, Ed.). Bishop Museum Press, Honolulu. *(256)*

Grime, J. P. (1974). Vegetation classification by reference to strategies. *Nature* **250**: 26–30. *(229)*

Grime, J. P. (1977). Evidence for the existence of three primary strategies in plants and its relevance to ecological and evolutionary theory. *Amer. Natur.* **111**: 1169–1194. *(229)*

Grinnell, J. (1917). The niche-relationships of the California thrasher. *Auk* **34**: 427–433. *(93)*

Guilday, J. E., P. S. Martin, and A. D. McCrady (1964). New Paris No 4: A late Pleistocene cave deposit in Bedford County, Pennsylvania. *Nat. Speleo. Soc. Bull.* **26**: 121–194. *(121,231)*

Haffer, J. (1969). Speciation in Amazonian forest birds. *Science* **165**: 131–137. *(127,128)*

Haldane, J. B. S. (1948). The theory of a cline. *J. Genetics* **48**: 277–284. *(305)*

Hall, O. (1972). Oxygen requirements of root meristems in diploid and autotetraploid rye. *Hereditas* **70**: 69–74. *(303)*

Hallam, A. (1973). Distribution patterns in contemporary terrestrial and marine animals. In *Organisms and Continents through Time* (N. F. Hughes, Ed.). The Palaeontological Society, London. Special papers. *(152)*

Haskell, G. (1952). Polyploidy, ecology and the British flora. *J. Ecol.* **40**: 265–282. *(300)*

Hayden, B. P. and R. Dolan (1976). Coastal marine fauna and marine climates of the Americas. *J. Biogeography* **3**: 71–81. *(21,161)*

Hays, J. D. (1967). Quaternary sediments in the Antarctic Ocean. *Progress Oceanogr.* **4**: 117–131. *(156)*

Heatwole, H. (1971). Marine-dependent terrestrial biotic communities on some cays in the Coral Sea. *Ecology* **52**: 363–366. *(183)*

Heatwole, H. and R. Levins (1972). Trophic structure stability and faunal change during recolonization. *Ecology* **53**: 531–534. *(182)*

Heatwole, H. and R. Levins (1973). Biogeography of the Puerto Rican Bank: species-turnover on a small cay, Cayo Ahogado. *Ecology* **54:** 1042–1055. *(181,182)*

Hecht, A. D. (1969). Miocene distribution of Molluskan provinces along the east coast of the United States. *Geol. Soc. Amer. Bull.* **80:** 1617–1620. *(161,162)*

Hecht, A. D. and B. Agan (1972). Diversity and age relationships in Recent and Miocene bivalves. *Syst. Zool.* **21:** 308–312. *(163)*

Hennig, W. (1965). Phylogenetic systematics. *Ann. Rev. Ent.* **10:** 97–116. *(66)*

Hennig, W. (1975). "Cladistic analysis or cladistic classification?" A reply to Ernst Mayr. *Syst. Zool.* **24:** 244–256. *(68,72)*

Hewer, H. R. (1971). Modern zoogeographical regions. In *Faunal Provinces in Space and Time.* (F. A. Middlemiss and P. F. Rawson, Eds.). Seel House Press, Liverpool. *(15)*

Hopkins, D. M. (1967). The Cenozoic history of Beringia—a synthesis. In *The Bering Land Bridge* (D. M. Hopkins, Ed.). Stanford University Press, Stanford. *(131,133,135)*

Hoppe, G. (1963). Some comments on the 'ice-free refugia' of northern Scandinavia. In *North Atlantic Biota and Their History* (A. Löve and D. Löve, Eds.). Pergamon Press, New York. *(120)*

Hosie, R. C. (1969). *Native Trees of Canada.* Canadian Forestry Service, Ottawa. *(219)*

Howard, R. A. (1960). *Dynamic Programming and Markov Processes.* Wiley, New York. *(103)*

Huey, R. B. (1978). Latitudinal pattern of between-altitude faunal similarity: mountains might be "higher" in the tropics. *Amer. Natur.* **112:** 225–229. *(228)*

Hultén, E. (1958). *The Amphi-Atlantic Plants and their Phytogeographical Connections.* Almquist and Wiksell, Stockholm. *(255,276)*

Humm, H. J. (1969). Distribution of marine algae along the Atlantic coast of North America. *Phycologia* **7:** 43–53. *(161)*

Hurley, P. M. (1968). The confirmation of continental drift. In *Continents Adrift* (J. T. Wilson, Ed.). Freeman, San Francisco. *(52)*

Hutchins, L. W. (1947). The bases for temperature zonation in geographical distribution. *Ecol. Monogr.* **17:** 325–335. *(161)*

Hutchinson, G. E. (1957). Concluding remarks. *Cold Spring Harbor Symp. Quant. Biol.* **22:** 415–427. *(93)*

Hutchinson, J. (1926). *The Families of Flowering Plants. I. Dicotyledons* Macmillan, London. *(132,268,271)*

Illies, J. (1965). Phylogeny and zoogeography of the Plecoptera. *Ann. Rev. Ent.* **10:** 117–140. *(79)*

Jackson, J. B. C. (1974). Biogeographic consequences of eurytopy and stenotopy among marine bivalves and their evolutionary significance. *Amer. Natur.* **108:** 541–560. *(166)*

Jardine, N. and D. McKenzie (1972). Continental drift and the dispersal and evolution of organisms. *Nature* **235:** 20–24. *(54)*

Jelinek, A. J. (1967). Man's role in the extinction of Pleistocene faunas. In *Pleistocene Extinctions. The Search for a Cause* (P. S. Martin and H. E. Wright Jr., Eds.). Yale University Press, New Haven. *(137)*

Johnson, A. W. and J. G. Packer (1965). Polyploidy and environment in arctic Alaska. *Science* **148:** 237–239. *(301)*

Johnson, M. P. and P. H. Raven (1973). Species numbers and endemism: the Galapagos Archipelago revisited. *Science* **179:** 893–895. *(193,194)*

Johnson, M. W. and E. Brinton (1963). Biological species, water masses and currents. In *The Sea*, Vol. 2 (M. N. Hill, Ed.). Interscience, New York. *(154,274)*

Keast, A. (1972). Continental drift and the biota of the mammals on southern continents. In *Evolution, Mammals and Southern Continents* (A. Keast, F. C. Erk, and B. Glass, Eds.). State University of New York Press, Albany. *(55)*

Keast, A. (1973). Contemporary biotas and the separation sequence of the southern continents. In *Implications of Continental Drift to the Earth Sciences*, Vol. 1 (D. H. Tarling and S. K. Runcorn, Eds.). Academic Press, New York. *(28,31,48)*

Keast, A., F. C. Erk, and B. Glass (1972). *Evolution, Mammals and Southern Continents*. State University of New York Press, Albany. *(v)*

Kennett, J. P. (1968). Latitudinal variation in *Globigerina pachyderma* (Ehrenberg) in surface sediments of the southwest Pacific Ocean. *Micropaleontology* **14:** 305–318. *(156)*

Kikkawa, J. and K. Pearse (1969). Geographical distribution of land birds in Australia—a numerical analysis. *Austr. J. Zool.* **17:** 821–840. *(17,20)*

Klekowski, E. J. Jr. (1972). Genetical features of ferns as contrasted to seed plants. *Ann. Missouri Bot. Gard.* **59:** 138–151. *(297)*

Kornas, J. (1972). Corresponding taxa and their ecological background in the forests of temperate Eurasia and North America. In *Taxonomy, Phytogeography and Evolution* (D. H. Valentine, Ed.). Academic Press, New York. *(38,204,222)*

Kowalski, K. (1967). The Pleistocene extinction of mammals in Europe. In *Pleistocene Extinctions. The Search for a Cause* (P. S. Martin and H. E. Wright Jr., Eds.). Yale University Press, New Haven. *(129)*

Kummel, B. (1970). *History of the Earth*. 2nd ed. Freeman, San Francisco. *(129)*

Kurtén, B. (1969). Continental drift and evolution. In *Continents Adrift* (J. T. Wilson, Ed.). Freeman, San Francisco. *(56,95)*

Kurtén, B. (1971). *The Age of Mammals*. Columbia University Press, New York. *(110)*

Kurtén, B. (1972). *The Ice Age*. Rupert Hart-Davis, London. *(56, 111,112,129,137)*

Lack, D. (1970). Island birds. *Biotropica* **2:** 29–31. *(176)*

Langford, A. N. and M. F. Buell (1969). Integration, identity and stability in the plant association. *Adv. Ecol. Res.* **6:** 83–135. *(217)*

Leopold, E. B. (1967). Late Cenozoic patterns of plant extinction. In *Pleistocene Extinctions. The Search for a Cause* (P. S. Martin and H. E. Wright Jr., Eds.). Yale University Press, New Haven. *(41)*

Le Pichon, X. (1968). Sea-floor spreading and continental drift. *J. Geophys. Res.* **73:** 3661–3697. *(27,28)*

Li, H. L. (1952). Floristic relationships between eastern Asia and eastern North America. *Trans. Amer. Phil. Soc.* New Series **42:** 371–429. *(132,269)*

Lindroth, C. H. (1957). *The Faunal Connections between Europe and North America*. Almquist and Wiksell, Stockholm. *(278)*

Lindroth, C. H. (1960). Is Davis Strait—between Greenland and Baffin Island—a floristic barrier? *Botaniska Notiser* **113(2):** 129–140. *(252)*

Lindroth, C. H. (1963a). The problem of late land connections in the North Atlantic area. In *Biota of the North Atlantic and Their History* (A. Löve and D. Löve, Eds.). Pergamon Press, New York. *(253)*

Lindroth, C. H. (1963b). The Aleutian Islands as a route for dispersal across the North Pacific. In *Pacific Basin Biogeography* (J. L. Gressitt, Ed.). Bishop Museum Press, Honolulu. *(245)*

Linsley, E. G. (1963). Bering Arc relationships of Cerambycidae and their host plants. In *Pacific Basin Biogeography* (J. L. Gressitt, Ed.). Bishop Museum Press, Honolulu. *(132,270)*

Llano, G. A. (1965). The flora of Antarctica. In *Antarctica* (T. Hatherton, Ed.). New Zealand Antarctic Society Survey. Frederick A. Praeger, New York. *(43)*

Löve, A. (1958). Transatlantic connections and long-distance dispersal. *Evolution* **12**: 421–423. *(252)*

Löve, D. (1963). Dispersal and survival of plants. In *North Atlantic Biota and Their History* (A. Löve and D. Löve, Eds.). Pergamon Press, New York. *(263)*

Lucas, J. (1974). *Life in the Oceans*. Dutton, New York. *(168)*

Lynch, J. F. and N. K. Johnson (1974). Turnover and equilibria in insular avifaunas, with special reference to the California Channel Islands. *Condor* **76**: 370–384. *(181)*

MacArthur, R. H. and E. O. Wilson (1963). An equilibrium theory of insular biogeography. *Evolution* **17**: 373–387. *(2,174)*

MacArthur, R. H. and E. O. Wilson (1967). *The Theory of Island Biogeography*. Princeton University Press, Princeton. *(174,186)*

McDowall, R. M. (1973). Zoogeography and taxonomy. *Tuatara* **20**: 88–96. *(61)*

McGowan, J. A. (1963). Geographical variation in *Limacina helicina* in the North Pacific. In *Speciation in the Sea* (J. P. Harding and N. Tebble, Eds.). The Systematic Association, London, Publication No. 5. *(171)*

McGowan, J. A. (1971). Oceanic biogeography of the Pacific. In *The Micropalaeontology of Oceans* (B. M. Funnell and W. R. Riedel, Eds.). Cambridge University Press, Cambridge. *(140,141, 153,275)*

McIntosh, R. P. (1967). The continuum concept of vegetation. *Bot. Rev.* **33**: 130–187. *(217)*

McKenna, M. C. (1975). Fossil mammals and early Eocene North Atlantic continuity. *Ann. Missouri Bot. Gard.* **62**: 335–353. *(31,32,249)*

McLean, D. M. (1978). Land floras: the major late Phanerozoic atmospheric carbon dioxide/oxygen control. *Science* **200**: 1060–1062. *(139)*

Macpherson, A. H. (1968). Land mammals. In *Science, History and Hudson Bay*, Vol. 1 (C. S. Beals, Ed.). Department of Energy, Mines and Resources, Ottawa. *(126)*

Malfait, B. T. and M. G. Dinkelman (1972). Circum-Caribbean tectonic and igneous activity in the evolution of the Caribbean plate. *Geol. Soc. Amer. Bull.* **83**: 251–272. *(32,34)*

Marcus, L. F. and J. H. Vandermeer (1966). Regional trends in geographic variation. *Syst. Zool.* **15**: 1–13. *(233)*

Margolis, S. V. and J. P. Kennett (1971). Cenozoic paleoglacial history of Antarctica recorded in Subantarctic deep sea cores. *Amer. J. Sci.* **271**: 1–36. *(44)*

Marshall, N. B. (1963). Diversity, distribution and speciation of deep-sea fishes. In *Speciation in the Sea* (J. P. Harding and N. Tebble, Eds.). The Systematics Society, London, Publication No. 5. *(168)*

Martin, P. S. (1967). Prehistoric overkill. In *Pleistocene Extinctions. The Search for a Cause* (P. S. Martin and H. E. Wright Jr., Eds.). Yale University Press, New Haven. *(129)*

Martinsson, A. (1975). Editor's column: Planktic, nektic, benthic. *Lethaia* **8**: 193–194. *(144n)*

Mason, H. L. (1954). Migration and evolution in plants. *Madroño* **12**: 161–192. *(243)*

Mattingly, P. F. (1962). Towards a zoogeography of the mosquitoes. In *Taxonomy and Geography* (D. Nichols, Ed.). The Systematics Association, London, Publication No. 4. *(14)*

May, R. M., J. A. Endler, and R. E. McMurtrie (1975). Gene frequency clines in the presence of selection opposed by gene flow. *Amer. Natur.* **109**: 659–676. *(83,305,310)*

Mayer, A. G. (1910). *Medusae of the World. III The Scyphomedusae.* Carnegie Institute of Washington, Washington, D.C. *(279)*

Mayr, E. (1965). Avifauna: turnover on islands. *Science* **150**: 1587–1588. *(193)*

Mayr, E. (1967). Evolutionary challenges to the mathematical interpretation of evolution. *Wistar Inst. Symp. Monogr.* No. 5: 47–58. *(88)*

Mayr, E. (1970). *Population, Species, and Evolution.* Harvard University Press, Cambridge, Mass. *(82,192)*

Melville, R. (1973). Continental drift and plant distribution. In *Implications of Continental Drift to the Earth Sciences*, Vol. 1 (D. H. Tarling and S. K. Runcorn, Eds.). Academic Press, New York. *(40)*

Menzies, R. J., R. Y. George, and G. T. Rowe (1973). *Abyssal Environment and Ecology of the World Oceans.* Wiley, New York. *(144,148,150)*

Middlemiss, F. A. and P. F. Rawson (1971). Faunal provinces in space and time—some general considerations. In *Faunal Provinces in Space and Time* (F. A. Middlemiss and P. F. Rawson Eds.). Seel House Press, Liverpool. *(151)*

Mirov, N. T. (1967). *The genus Pinus*. Ronald Press, New York. *(218)*

Mosquin, T. (1967). Evidence for autopolyploidy in *Epilobium angustifolium* (Onagraceae). *Evolution* **21**: 713–719. *(303)*

Müller, D. G. (1964). Life-cycle of *Ectocarpus siliculosus* from Naples, Italy. *Nature* **203**: 1402. *(295)*

Murray, J. W. (1973). *Distribution and Ecology of Living Benthic Foraminiferids*. Heinemann, London. *(154)*

Mutch, R. W. (1970). Wildland fires and ecosystems—a hypothesis. *Ecology* **51**: 1046–1051. *(212)*

Nannfeldt, J. A. (1963). Taxonomic differentiation as an indicator of the migratory history of the North Atlantic flora with especial regard to the Scandes. In *North Atlantic Biota and Their History* (A. Löve and D. Löve, Eds.). Pergamon Press, New York. *(303)*

Neumann, G. and W. J. Pierson Jr. (1966). *Principles of Physical Oceanography*. Prentice-Hall, Englewood Cliffs, N.J. *(142,153, 159)*

Omodeo, P. (1963). Distribution of the terricolous oligochaetes on the two shores of the Atlantic. In *North Atlantic Biota and Their History* (A. Löve and D. Löve, Eds.). Pergamon Press, New York. *(246,265)*

Papenfuss, G. F. (1935). Alternation of generations in *Ectocarpus siliculosus*. *Bot. Gazette* **96**: 421–446. *(295)*

Parsons, R. F. (1969). Distribution and paleogeography of two mallee species of *Eucalyptus* in southern Australia. *Austr. J. Bot.* **17**: 323–330. *(112)*

Peattie, D. C. (1922). The Atlantic coastal plain element in the flora of the Great Lakes. *Rhodora* **24**: 57–70; 80–88. *(113)*

Phleger, F. B. (1960). *Ecology and Distribution of Recent Foraminifera*. Johns Hopkins University Press, Baltimore. *(155)*

Pickard, G. L. (1975). *Descriptive Physical Oceanography*. 2nd ed., Pergamon Press, New York. *(141)*

Pielou, E. C. (1973). Geographic variation in host-parasite specificity. In *The Mathematical Theory of the Dynamics of Biological Populations* (M. S. Bartlett and R. W. Hiorns, Eds.). Academic Press, New York. *(293)*

Pielou, E. C. (1974a). Biogeographic range comparisons and evidence of geographic variation in host-parasite relations. *Ecology* **55**: 1359–1367. *(290n,291)*

Pielou, E. C. (1974b). *Population and Community Ecology*. Gordon and Breach, New York. *(227n)*

Pielou, E. C. (1975). *Ecological Diversity.* Wiley, New York. *(93,95,223,228)*

Pielou, E. C. (1977a). *Mathematical Ecology.* Wiley, New York. *(17,92,99,217,307)*

Pielou, E. C. (1977b). The latitudinal spans of seaweed species and their patterns of overlap. *J. Biogeography* **4:** 299–311. *(21,99)*

Pielou, E. C. (1978a). The statistics of biogeographic range maps: sheaves of one-dimensional ranges. *Proc. 41st. Congr. Intern. Statist. Inst.,* Delhi. *(99,165)*

Pielou, E. C. (1978b). Latitudinal overlap of seaweed species: evidence for quasi-sympatric speciation. *J. Biogeography* **5:** 227–238. *(21,84,99,165,225)*

Platnick, N. I. (1976). Drifting spiders or continents? Vicariance biogeography of the spider subfamily Laroniinae (Araneae: Gnaphosidae). *Syst. Zool.* **25:** 101–109. *(63)*

Plumstead, E. P. (1973). The enigmatic *Glossopteris* flora and uniformitarianism. In *Implications of Continental Drift to the Earth Sciences*, Vol. 1 (D. H. Tarling and S. K. Runcorn, Eds.). Academic Press, New York. *(45)*

Poole, A. L. (1950). Studies of New Zealand *Nothofagus.* I Taxonomy and floral morphology. *Trans. Roy. Soc. N.Z.* **78:** 363–380. *(247)*

Porsild, A. E. (1957). *Illustrated Flora of the Canadian Arctic Archipelago.* National Museums of Canada, Ottawa. Bull. No. 146. *(255)*

Porsild, A. E. (1958). Geographical distribution of some elements in the flora of Canada. *Geog. Bull.* **11:** 57–77. *(121)*

Porsild, A. E. (1969). Discussion of Drury's paper in *Essays in Plant Geography and Ecology* (K. N. H. Greenidge, Ed.). Nova Scotia Museum, Halifax. *(117)*

Porter, D. M. (1976). Geography and dispersal of Galapagos Island vascular plants. *Nature* **264:** 745–746. *(260)*

Preest, D. S. (1963). A note on the dispersal characteristics of the seed of the New Zealand podocarps and beeches and their biogeographical significance. In *Pacific Basin Biogeography* (J. L. Gressitt, Ed.). Bishop Museum Press, Honolulu. *(246)*

Preston, F. W. (1962). The canonical distribution of commonness and rarity. *Ecology* **43:** 185–215; 410–432. *(174)*

Proctor, V. W., C. R. Malone, and V. L. DeVlaming (1967). Dispersal of aquatic organisms: viability of disseminules recovered from the intestinal tract of captive killdeer. *Ecology* **48:** 672–676. *(263)*

Raup, D. M. (1972). Taxonomic diversity during the Phanerozoic. *Science* **177:** 1065–1071. *(95)*

Raven, P. H. (1963). Amphitropical relationships in the floras of North and South America. *Quart. Rev. Biol.* **38:** 151–177. *(271,272, 275,277)*

Raven, P. H. and D. I. Axelrod (1972). Plate tectonics and Australasian paleobiogeography. *Science* **176:** 1379. *(49,248)*

Raven, P. H. and D. I. Axelrod (1974). Angiosperm biogeography and past continental movements. *Ann. Miss. Bot. Gard.* **61:** 539–673. *(28,41,43,49,52,131,151,209,249,252)*

Rehn, J. A. G. (1958). The origin and affinities of the Dermaptera and Orthoptera of western North America. In *Zoogeography* (C. L. Hubbs, Ed.). AAAS, Washington, D.C. *(277)*

Reichle, D. E. (1966). Some Pselaphid beetles with boreal affinities and their distribution along the postglacial fringe. *Syst. Zool.* **15:** 330–344. *(122)*

Repenning, C. A. (1967). Palearctic-Nearctic mammalian dispersal in the late Cenozoic. In *The Bering Land Bridge* (D. M. Hopkins, Ed.). Stanford University Press, Stanford. *(136)*

Ricklefs, R. E. and G. W. Cox (1972). Taxon cycles in the West Indian avifauna. *Amer. Natur.* **106:** 195–219. *(195)*

Ridley, H. N. (1930). *The Dispersal of Plants Throughout the World.* L. Reeve and Co. Ashford, Kent, England. *(259,261)*

Roland, A. E. and E. C. Smith (1969). *The Flora of Nova Scotia.* Nova Scotia Museum, Halifax. *(220)*

Rosen, D. E. (1975). The vicariance model of Caribbean biogeography. *Syst. Zool.* **24:** 431–464. *(34,79n,80)*

Ross, H. H. (1972). An uncertainty principle in ecological evolution. In *A Symposium on Ecosystematics* (R. T. Allen and F. C. James, Eds.). University of Arkansas, Fayetteville. *(63)*

Rowe, J. S. and G. W. Scotter (1973). Fire in the boreal forest. *Quaternary Research* **3:** 444–464. *(213,214)*

Sauer, J. D. (1969). Oceanic islands and biogeographic theory: a review. *Geog. Rev.* **59:** 582–593. *(176,265)*

Savile, D. B. O. (1956). Known dispersal rates and migratory potentials as clues to the origin of the North American biota. *Amer. Midland Naturalist* **56:** 434–453. *(253)*

Savile, D. B. O. (1968). Land Plants. In *Science, History and Hudson Bay*, Vol 1 (C. S. Beals, Ed.). Department of Energy, Mines and Resources, Ottawa. *(124,126,261)*

Scheltema, R. S. (1971). Larval dispersal as a means of genetic exchange between geographically separated populations of shallow-water benthic marine gastropods. *Biol. Bull. Woods Hole* **140**: 284–322. *(170,278)*

Scheltema, R. S. (1972). Dispersal of larvae as a means of genetic exchange between widely separated populations of shoal water benthic invertebrate species. In *Fifth European Marine Biology Symposium* (B. Battaglia, Ed.). Piccin Editore, Padova, Italy. *(170)*

Schlinger, E. I. (1974). Continental drift, *Nothofagus*, and some ecologically associated insects. *Ann. Rev. Ent.* **19**: 323–343. *(247)*

Schmidt, K. P. (1954). Faunal realms, regions and provinces. *Quart. Rev. Biol.* **29**: 322–331. *(5)*

Schminke, H. K. (1974). Mesozoic intercontinental relationships as evidenced by bathynellid Crustacea (Syncarida: Malacostraca). *Syst. Zool.* **23**: 157–164. *(60,79)*

Schuster, R. M. (1976). Plate tectonics and its bearing on the geographical origin and dispersal of the angiosperms. In *The Origin and Early Evolution of Angiosperms* (C. B. Beck, Ed.). Columbia University Press, New York. *(42)*

Scott, G. H. (1976). Foraminiferal biostratigraphy and evolutionary models. *Syst. Zool.* **25**: 78–80. *(88)*

Seddon, G. (1974). Xerophytes, xeromorphs and sclerophylls: the history of some concepts in ecology. *Biol. J. Linn. Soc.* **6**: 65–87. *(211n,213)*

Simberloff, D. S. (1969). Experimental zoogeography of islands. A model for insular colonization. *Ecology* **50**: 296–314. *(181)*

Simberloff, D. S. (1970). Taxonomic diversity of island biotas. *Evolution* **24**: 23–47. *(225)*

Simberloff, D. S. (1974). Equilibrium theory of island biogeography and ecology. *Ann. Rev. Ecol. Syst.* **5**: 161–182. *(176,180,182)*

Simpson, G. G. (1947). Evolution, interchange and resemblance of the North American and Eurasian Cenozoic mammalian faunas. *Evolution* **1**: 218–220. *(251)*

Simpson, G. G. (1950). History of the fauna of Latin America. *Amer. Scientist* **38**: 361–389. *(57,227)*

Simpson, G. G. (1952). Probabilities of dispersal in geologic time. *Bull. Amer. Mus. Nat. Hist.* **99:** 163–176. *(264)*

Simpson G. G. (1953). *The Major Features of Evolution.* Columbia University Press, New York. *(v)*

Simpson, G. G. (1969). The first three billion years of community evolution. In *Diversity and Stability in Ecological Systems* (G. M. Woodwell and H. H. Smith, Eds.). Brookhaven National Laboratory, U.S. Atomic Energy Commission, Brookhaven. *(57)*

Sjörs, H. (1963). Amphi-Atlantic zonation, Nemoral to Arctic. In *North Atlantic Biota and Their History* (A. Löve and D. Löve, Eds.). Pergamon Press, New York. *(216,220)*

Slatkin, M. (1973). Gene flow and selection in a cline. *Genetics* **75:** 733–756. *(83,305,310)*

Sokal, R. R. (1975). Mayr on cladism—and his critics. *Syst. Zool.* **24:** 257–262. *(73)*

Solbrig, O. T. (1972). The floristic disjunctions between the "Monte" in Argentina and the "Sonoran Desert" in Mexico and the United States. *Ann. Miss. Bot. Gard.* **59:** 218–223. *(273)*

Specht, R. L. (1969). A comparison of the sclerophyllous vegetation characteristic of Mediterranean type climates in France, California and southern Australia. *Austr. J. Bot.* **17:** 277–292. *(209)*

Stanley, S. M. (1975). A theory of evolution above the species level. *Proc. Nat. Acad. Sci. U.S.* **72:** 646–650. *(88,89)*

Stebbins, G. L. (1971). *Chromosomal Evolution in Higher Plants.* Edward Arnold, London. *(297,299,301,304)*

Stegmann, B. (1963). The problem of Beringian continental land connection in the light of ornithogeography. In *Pacific Basin Biogeography* (J. L. Gressitt, Ed.). Bishop Museum Press, Honolulu. *(42,133)*

Stehli, F. G. (1968). Taxonomic gradients in pole location: the Recent model. In *Evolution and Environment* (E. T. Drake, Ed.). Yale University Press, New Haven. *(74)*

Stehli, F. G., R. G. Douglas, and N. D. Newell (1969). Generation and maintenance of gradients in taxonomic diversity. *Science* **164:** 947–950. *(167)*

Steindorsson, S. (1963). Ice age refugia in Iceland as indicated by the present distribution of plant species. In *North Atlantic Biota and Their History* (A. Löve and D. Löve, Eds.). Pergamon Press, New York. *(251)*

Stokes, W. L. (1973). *Essentials of Earth History.* 3rd. ed. Prentice-Hall, Englewood Cliffs, N.J. *(219)*

Strang, R. (1972). Succession in unburned subarctic woodlands. *Can. J. Forest Res.* **3:** 140–143. *(214)*

Suomalainen, E. (1961). On morphological differences and evolution of different polyploid parthenogenetic weevil populations. *Hereditas* **47:** 309–341. *(304)*

Suomalainen, E. (1962). Significance of parthenogenesis in the evolution of insects. *Ann. Rev. Ent.* **7:** 349–366. *(304)*

Sverdrup, H. V., M. W. Johnson, and R. H. Fleming (1942). *The Oceans: Their Physics, Chemistry and General Biology.* Prentice-Hall, New York. *(164)*

Sylvester-Bradley, P. C. (1971). Dynamic factors in animal palaeogeography. In *Faunal Provinces in Space and Time* (F. A. Middlemiss and P. F. Rawson, Eds.). Seel House Press, Liverpool. *(5,140,148)*

Szalay, F. S. (1977). Ancestors, descendents, sister groups and testing of phylogenetic hypotheses. *Syst. Zool.* **26:** 1–11. *(72)*

Takhtajan, A. (1969). *Flowering Plant Origin and Dispersal.* Smithsonian Institution Press, Washington, D.C. *(40)*

Taylor, J. D. and C. N. Taylor (1977). Latitudinal distribution of predatory gastropods on the eastern Atlantic shelf. *J. Biogeography* **4:** 73–81. *(167)*

Tedford, R. H. (1974). Marsupials and the new paleogeography. In *Paleogeographic Provinces and Provinciality* (C. A. Ross, Ed.). Soc. Economic Paleontologists and Mineralogists Spec. Publ. No. 21. *(29,54,55,151)*

Thomson, C. W. and J. Murray (1885). *Report of the Scientific Results of the Voyage of H. M. S. Challenger during the Years 1873–76. Zoology—Vol XIII.* Johnson Reprint Co., New York, 1965. *(274)*

Thorne, R. F. (1972). Major disjunctions in the geographic ranges of seed plants. *Quart. Rev. Biol.* **47:** 365–411. *(267)*

Valentine, J. W. (1966). Numerical analysis of marine molluscan ranges on the extratropical northeastern Pacific shelf. *Limnol. Oceanogr.* **11:** 198–211. *(24)*

Valentine, J. W. (1973). *Evolutionary Paleoecology of the Marine Biosphere.* Prentice-Hall, Englewood Cliffs, N.J. *(v)*

Valentine, J. W. and E. M. Moores (1972). Global tectonics and the fossil record. *J. Geol.* **80:** 167–184. *(95)*

Van den Hoek, C. (1975). Phytogeographic provinces along the coasts of the northern Atlantic Ocean. *Phycologia* **14:** 317–330. *(151,161)*

Vandermeer, J. H. (1966). Statistical analysis of geographic variation of the Fathead Minnow, *Pimephales promelas*. *Copeia* **1966(3):** 457–466. *(233)*

Vinogradova, N. G. (1959). The zoogeographical distribution of the deep-water bottom fauna in the abyssal zone of the ocean. *Deep-Sea Res.* **5:** 205–208. *(148,149)*

Vinogradova, N. G. (1962). Vertical zonation in the distribution of deep-sea benthic fauna in the ocean. *Deep-Sea Res.* **8:** 245–250. *(147)*

Von Arx, W. S. (1962). *An Introduction to Physical Oceanography.* Addison-Wesley, Reading, Mass. *(143)*

Vuilleumier, B. S. (1971). Pleistocene changes in the flora of South America. *Science* **173:** 771–780. *(127,128)*

Wagner, W. H. Jr. (1972). Disjunctions in homosporous vascular plants. *Ann. Miss. Bot. Gard.* **59:** 203–217. *(296,298)*

Wagner, W. H. Jr. and A. J. Sharp (1963). A remarkably reduced vascular plant in the United States. *Science* **142:** 1483–1484. *(296)*

Walker, E. P. (1968). *Mammals of the World.* 2nd ed. Johns Hopkins University Press, Baltimore. *(223)*

Watts, D. (1971). *Principles of Biogeography.* McGraw-Hill, New York. *(v)*

Webb, S. D. (1976). Mammalian faunal dynamics of the great American interchange. *Paleobiology* **2:** 220–234. *(57)*

Went, F. W. (1971). Parallel evolution. *Taxon* **20:** 197–226. *(214)*

Whitehead, D. R. (1972). Approaches to disjunct populations: the contribution of palynology. *Ann. Miss. Bot. Gard.* **59:** 125–137. *(120)*

Whitehead, D. R. and C. E. Jones (1969). Small islands and the equilibrium theory of insular biogeography. *Evolution* **23:** 171–179. *(178)*

Whitmore, T. C. (1973). Plate tectonics and some aspects of Pacific plant geography. *New Phytol.* **72:** 1185–1190. *(257)*

Wilcox, B. A. (1978). Supersaturated island faunas: a species-age relationship for lizards on post-Pleistocene landbridge islands. *Science* **199:** 996–998. *(191)*

Wilks, S. S. (1962). *Mathematical Statistics.* Wiley, New York. *(285)*

Williams, J. (1962). *Oceanography.* Little, Brown and Co., Boston. *(142)*

Williams, W. T. (1971). Principles of clustering. *Ann. Rev. Ecol. Syst.* **2:** 303–326. *(17)*

Wilson, E. O. (1961). The nature of the taxon cycle in the Melanesian ant fauna. *Amer. Natur.* **95**: 169–193. *(195)*

Wilson, J. T. (1963a). Continental drift. *Sci. American* **208(4)**: 86–100. *(2,27,34)*

Wilson, J. T. (1963b) Evidence from islands of the spreading of ocean floors. *Nature* **197**: 536–538. *(200,257)*

Wilson, J. T. (1965). Evidence from ocean islands suggesting movement in the earth. *Phil. Trans. Roy. Soc. A.* **258**: 145–167. *(200,257)*

Wolfe, J. A. (1971). Tertiary climatic fluctuations and methods of analysis of Tertiary floras. *Paleogeography, Paleoclimatology, Paleoecology* **9**: 27–57. *(44)*

Wolfe, J. A. and E. B. Leopold (1967). Neogene and early Quaternary vegetation of northwestern North America and northeastern Asia. In *The Bering Land Bridge* (D. M. Hopkins, Ed.). Stanford University Press, Stanford. *(131)*

Wolff, T. (1970). The concept of hadal or ultra-abyssal fauna. *Deep-Sea Res.* **17**: 983–1003. *(147)*

Wood, C. E. Jr. (1972). Morphology and phytogeography: the classical approach to the study of disjunctions. *Ann. Miss. Bot. Gard.* **59**: 107–124. *(269)*

Woodruff, D. S. (1978). Mechanisms of speciation. *Science* **199**: 1329–1330. *(83)*

Wright, S. (1967). Comments on the preliminary working papers of Eden and Waddington. *Wistar Inst. Symp. Monogr. No. 5:* 117–120. *(87)*

Zeeman, E. C. (1976). Catastrophe theory. *Sci. American* **234(4)**: 65–83. *(85)*

TAXONOMIC INDEX

(The species within a genus are not indexed separately)

SUBJECT INDEX

(Page numbers in bold face refer to chapter and section headings)